2013
YEARBOOK

THE CENTRE FOR FORTEAN ZOOLOGY

www.cfz.org.uk

Edited and Compiled by

Jonathan and Corinna Downes

Typeset by Jonathan Downes,
Cover and Layout by SPiderKaT for CFZ Communications
Using Microsoft Word 2000, Microsoft Publisher 2000, Adobe Photoshop CS.

First published in Great Britain by CFZ Press

CFZ Press
Myrtle Cottage
Woolsery
Bideford
North Devon
EX39 5QR

ISBN: 978-1-909488-06-9

Dedicated with love and gratitude to:

Steve Jones
Terry Colvin

(but for whom this volume would not exist)

Contents

More Cryptozoology

A listing of certain animals unrecognised by science or of a paranormal nature

By

Ronan Coghlan

Introduction

The purpose of this volume[*] is supplementary to my previous output on mystery animals. However, I have included more detailed references to sources of information, as much of the material between these covers has come to light since the publications of the standard works by Eberhardt and Newman.

If an article says that such and such a cryptid was seen, heard or otherwise apprehended, the reader should understand the word "allegedly" going before it, even if the word does not appear there.

It is by no means impossible that in some of the entries allegations are misperceptions or downright hoaxes. A book such as this is only as good as its sources. Sometimes, the sources can be checked out, but often not.

Generally speaking, animals in my *Dictionary of Cryptozoology, Cryptosup* and *Further Cryptozoology* are not included in this volume. There are some exceptions, where new information may have warranted them.

Alien Big Cats

The only Alien Big Cats I have included are those which are neither brown nor black. It is true these animals are the subjects of numerous reports, but I believe most of those in Europe and Australia can be explained in one of the following ways:-
(a) they are deliberately released animals or their descendants;
(b) they are very large, perhaps mutant, domestic cats;
(c) they are the result of misperception or, perhaps in a few cases, downright hoaxes

---

[*] EDITOR'S NOTE: This was originally intended as another volume in Xiphos Books' highly acclaimed 'Dictionary of Cryptozoology' series. We are very grateful to Ronan for giving it to us instead.

In the United States, the situation is somewhat different. Some of the brown cats may be populations of puma which have returned to territories whence their ancestors were driven, such as now abandoned farmland. There are also many reports of big black cats in America. Although it has been suggested these are black pumas, the existence of black pumas has never been definitely established to the satisfaction of zoologists. While I suspect some of these may be released animals or their descendants, it should be borne in mind that traditions of large black cats, different from pumas, exist in Native American tradition. These were called Devil Cats. There was also a black felid called the Wooleneag, supposedly living in Maine in early times. Perhaps isolated populations of these creatures still exist. It is also possible that there are jaguarundis, which are black and feline, in the United States or that the sightings are misperceptions of the fisher, a kind of marten. The odd black bear might be mistaken for a black panther. One may find jaguars in Arizona these days and jaguars can sometimes be black, but current records do not show any black ones north of Panama. In all, I feel that big cat reports generally are to be explained by one of the above suggestions and that to list them all would result in a large and repetitious work.

Chupacabras

The chupacabras ("goat-sucker") was first reported from Puerto Rico. However, due to a kind of media frenzy, the term chupacabras has been applied to all sorts of animals in the Hispanic world and other areas. I suspect the original animals were only found in Puerto Rico and a plausible explanation of what they were may be found in Jonathan Downes' book *Island of Paradise.* The memory of the original witness may have been unconsciously influenced by the monster in the film *Species.* When you read of animals such as the Dry Gulch chupacabras in this work, the present writer does not feel we are dealing with a variety of the Puerto Rican animal. As to the name of this creature, the original form was *chupacabras*, but the term *chupacabra*, a more regular Spanish form, is also frequently found.

Paranormal

Many cryptozoologist will be distressed by my introduction of certain paranormal topics into a work on cryptozoology, because they would hold that it retards its acceptance into the realm of science. However, the paranormal must be understood as those elements which may exist but of which science is as yet unaware, just as, in days agone, gravity and electricity lay beyond the ken of our ancestors. There is no doubt that, in the case of some mystery animals, there is a paranormal aspect and, just because it cannot be scientifically analysed on the basis of our present knowledge, it does not mean we should avoid all consideration of it. We do not know the limits of the possible, however much certain persons may think we do.

In the case of cryptozoology, the following paranormal areas are relevant:-

When animals which could not be supported by their environment are reported by trustworthy witnesses, allowing for the possibility of error on the part of those witnesses, we must consider the possibility of alternative universes which are sometimes or always sufficiently conjoined to ours to allow them to slip through. Such considerations do not fly in the face of all modern speculation, it might be added. Some mainstream scientists seem to admit the possibility of alternative universes without commenting on the possibility of access to and from them. Indeed, Professor Michio Kaku of New York has suggested that when this universe wears out, we might have the ability to move to another. The same argument might be applied to isolated animals, of which only a single specimen has ever been reported, particularly if the specimen has characteristics which strike us as bizarre. Such creatures may have found a wormhole through which they have reached this universe.

In dealing with the latter kind of animal, we may wonder if we know about all that goes on in genetic research laboratories and whether creatures engendered by experimentation may not at times become fugitives from such places, adding to our lists of cryptids.

We might also consider the question of vibrations. It is perhaps possible that creatures vibrate at such a rate that makes them undiscernable to us for most of the time, but may occasionally vary their mode of vibration, making them occasionally apprehensible to our consciousness.

I am neither a believer nor disbeliever in UFOs, but, if they exist, wheresoever they come from - outer space, other dimensions, inside the earth or beneath the sea - it is possible they drop off from time to time creatures amenable to the earth's atmosphere and environment, but undocumented by zoologists.

Lastly, I might mention the argument put forward by some cryptozoologists that certain cryptids are tulpas. A tulpa is a thought form which a magician first visualises and on which he then confers an outward reality. Tulpas are held to be sentient beings, perhaps with a consciousness. They are to be found in Tibetan Buddhist belief. However, their creation would entail such willpower that it should be a possibility approached with caution, if it is a possibility at all. Alexandra David-Neel (1868-1969), a well-known traveller in the orient, claimed to have created a tulpa which in due course developed an independence of character which led her to eliminate him, a task accomplished with some difficulty.

After these considerations, I feel I am justified in including certain paranormal elements in this work, nor have I eschewed beasts from mythology if they might have some prototype in reality.

Finally, most of the creatures listed herein do not occur in my previous works on the subject. These are readily available on Amazon or from the publishers. This little opus may be regarded as in many respects a continuation of those works. My *Dictionary of Cryptozoology* (2004) deals with all the major cryptids, such as the Loch Ness Monster, Yeti, Bigfoot, etc., not to mention a host of lesser known creatures. Most of the creatures in this book are on the obscure side. Some are perhaps known only from single reports.

A

ADDISON BIRD A huge unidentified bird seen over Webster Springs (West Virginia), then called Addison, in 1895. It may have carried off a child and was seen carrying off a fawn. *MAWV:* 22

AFGHAN MYSTERY CAT An unidentified species of cat has been reported from Afghanistan. Images of the cat have been seen on night vision and thermal imaging equipment. *FZ:* January 16th, 2012

AGTA A hominid supposedly found on the island of Cebu in the Philippines. It can exceed 7' in height. www.bigfootencounters.com

AIKANAKA Hairy giants said to live near the Waianao Range on Oahu (Hawaii). *TG:* 68

AKALPUIS   In the beliefs of the Guajiro Indians of northern South America, short hairy beings the colour of ash.
AE: 90

A-KE   A kind of alligator reported from Korea. It may be that the Chinese alligator is to be found in Korea, its presence hitherto undocumented.
ACC

AKKOROKAMUI   A large marine creature supposed to lurk in Funka Bay, Japan, by the Ainu, the aboriginal people of the country, who preceded the main Japanese population. It is perhaps tentacled, for it has been compared with an octopus or squid.
Wikipedia

ALICANTE   Legendary creature of the Sierra Norte, Spain. It appears to be a sort of reptile, lacking legs or having very short ones, generally hairy, blind and poisonous. Suggestions are that it is a snake or lizard. It has also been suggested that it is a sighting of mongooses walking in a line.
criptolzoología en españa: March 21st, 2012.

ALIEN POD   This unidentified creature was reported from an artificial lake in Newport News (Va). Although someone concluded it must be extraterrestrial, thus giving it its name, there is no evidence for this. It is about 4' in diameter. Scientists feel that, rather than being a single animal, it is a colony of *Pectinatella magnifica.*
www.cfz.org.uk: November 5th, 2010

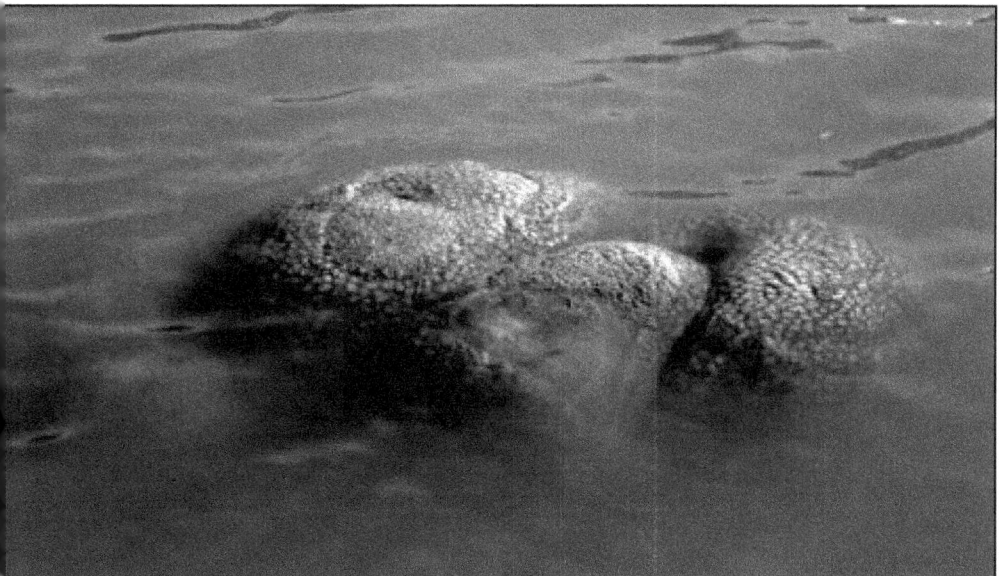

ALLENDE CREATURE A figure which appears to have been half human, half bird, with large wings, reported by a farmhand in Nuevo Leon, Mexico, in 1994.
Brian Gaugler: Things That Go Bump in the Night (website).

ALLIGATOR MAN In Colombian folklore, this creature (Spanish, *hombre cayman*), of which there may be only one, combines features of alligator and human.
Wikipedia

AL-MI'RAJ In Iranian legend, a single-horned rabbit.

ALPINE SERPENT This creature was allegedly discovered near Alpine (Utah) by people who moved a boulder to find a staircase or tunnel underneath. It led downwards. There they found a huge serpent from which they beat a hasty retreat.
Utah UFO Hunters Utah Creatures

AMARU A strange beast in Inca legend. It had the head of a llama, a fox's mouth, a condor's wings and a snake's body.
Wikipedia

AMBIRAK These creatures are now found in iconographic representation and are supposed to be female spirits occupying rivers. They feature in the myths of the Asmat in Irian Jaya, Indonesia. They may be based on actual creatures, such as pterosaurs.
FZ: July 17th, 2011

AMEMASU According to Ainu belief, this creature, which resembles a whale or a fish, is found in Lake Masu, Hokkaido, Japan.
Wikipedia

AMOMONGO In Filipino belief, a simian creature, man-sized and hirsute. It is sometimes called the Negros Ape, for it is to be found in West Negros, in caves at the base of Mount Kanlaon.
Wikipedia

ANCIENT ONE A creature in Estonian belief. It looks like a huge fish on four legs and its mouth, when opened, is terrifying to behold. It has a ridge on its back. L. Thomas says there was a reported sighting in 1987.
WW: 107

ANGHISTRI HUMANOIDS These are diaphanous creatures with membranes growing between their fingers. They are supposed to live off the coast of Anghistri, Greece.
HG: 70

ANIMAL SMALL PEOPLE Bipedal hirsute creatures in the folklore of the Takelma Indians of Oregon and California.
AE: 147

ANKLE BUG   In Berwick (Pa) in the year 1901 a strange epidemic of severe ankle bites was ascribed to this insect. It was never identified.
Masks of Messingw: May 2nd, 2012

ANTAMBA   A fierce beast resembling a felid said to be found in the Impenetrable Forest (Zahamena National Park) in Madagascar. It may be that the island still boasts a population of the supposedly extinct giant fossa (*Cryptoprocta spelea*).

ARALEZ   Creatures resembling dogs in Armenian myths. They had the power, by licking, to revive the dead.
www.armeniapedia.com

ARCADIAN GIANT   A giant over 5m tall observed by a witness in 1964.
JHS: 1:5

ARCADIAN SERPENT   In the 1860s it was said huge serpents were found in Arcadia in Greece. It was said by some that they were horned and roared.
HG: 58

ARCHBISHOP WALKENDORF'S BEAST   This prelate, Archbishop of Nidrosia (now Trondheim in Norway), sent the head of an unidentified beast to Pope Leo X in 1520. It had

red eyes and quills. The head was large and the body from which it had been severed was said to have been not much larger.
WW: 73

ARGENTINE BIPED   Large hairy biped reported from east of Rosario de la Frontera by two athletes and a gaucho.
meta-religion.com

ARIZONA ANIMAL   This creature, seen in 1888, was described as having the neck of a deer, the body of a dog and a bushy tail.   The witness was one W.T. Bumer, but some Mexicans had also reported such an animal.
V: 140

ARKAN SONNEY   Fairy pig, capable of altering its size, found on the Isle of Man.
Encyclopedia Mythica

ARKANSAS APE MAN   A creature variously described as both large and small.   It has been noted, not in the wilderness, but in residential areas of north-west Arkansas from 2003.
www.americanmonsters: December 22nd, 2009

ARKANSAS SNIPE   A huge insect believed to dwell at the bottom of rivers.   It is said to be large enough to eat a cow.
www.cfz.org.uk: 30th September, 2010

AROWHANA LIZARD   In 1898 an unidentified lizard, 5' in length, was discerned at this location in New Zealand.   A search party found footprints, which were photographed.
nzcryptozoology.ucoz.com

ARRE RIVER MONSTER   Although not reported for some time, a population of these serpentine creatures has been believed to exist in this Swiss river.   The Aar Gorge was supposed to be a good place to see them.
www.americanmonsters.com: August 22nd, 2010

ARVA ANIMAL   An animal killed near Arva, Ireland, in 1937.   It looked somewhat like a badger, somewhat like an otter, but was neither and was never identified.
MAI: 84

ASTOR MONSTER   A beast reported near Astor (Florida).   It was seen both in the water (at Lake Dexter) and going into the woods.   In the 1960s some fishermen, looking down into the Saint John's River, saw a beast the size of an elephant walking on the river bed.
FUW: 52

ATARPIAK   A small species of seal in Greenlandic lore.   It was said to be no bigger than a hand, whitish, with a blackish spot on each side.   The belief in the creature was noted by the 19th Century writer Robert Brown, who was advised it was probably a myth.
BR: 15

ATOE PANDAK, ATOE RIMBA Alternative names for the orang pendek.

AUBREY CREATURE This was reported in early 2008. It was a humanoid, about 5' tall, it had long arms and was grey-white in colour.
www.iraap.org: rosales

AVON CREATURE An animal reported at Avon (Wisconsin) in 1891. It was said to be grey with black spots or stripes. Its trails were like those of a bear.
V: 632

AYA NAPA SEA MONSTER A sea-monster that has been reported off the coast of Cyprus, but which has never shown any hostility to fishermen. It is supposed to be frequently seen in the region of Cavo Greko.
Wikipedia

AYAYEMA In the beliefs of the Alakalaf Indians of Tierra del Fuego, a monstrous creature living in a swamp that emerges at night seeking prey.
PM

AZOV MONSTER The Sea of Azov is a small body of water that lies off the Crimea, to the north of the Black Sea. A strange animal was allegedly caught by fishermen here. The fishermen themselves hailed from Rostov. The creature moved like a human and rotated its eyeballs. It weighed about 200 lbs.
www,answerbag.com: January 1st, 2008

B

BACA CREATURE At Baca, Mexico, a group on the lookout for UFOs one night was startled, first by a pair of red eyes, then a sound of galloping, then the sound of flying wings. One of them discerned the wings disappearing into bushes. The people concerned do not seem clear about whether all the sounds were made by the same creature. The incident was reported in 2008.
www.iraap.com: rosales

BACHMAN'S WARBLER (right) A possibly extinct bird (*Vermivora bachmanii*), but it may still exist. There was a possible sighting in Cuba in 2002.

BACKHASTEN A white river-dwelling horse in Scandinavian folklore. Mount it and it will jump into the river with you.
Wikipedia

BADALISC Although I suspect the name of this animal is based on that of the *basilisk*, this is supposed to be a friendly animal found in the folklore of the north of Italy. It is supposed to live in the woods near Anderista. Every year there is a festival in which a man dressed up (pictured right by Luca Giarelli) as it is captured and regales people with gossip.
Italian Wikipedia

BAHKAUV Creature of German folklore, a giant calf, believed in in the region of Aachen. However, it is regarded as scaly, a somewhat reptilian characteristic. One of those who allegedly fought with this creature is Pepin the Short, king of the Franks from 752-8.
German Wikipedia

BAIJI The Chinese river dolphin (*Lipotes vexillifer*) is thought to have become extinct in 2006, though one may have been sighted in 2007. This means there is a possibility of its continued survival.
Wikipedia

BALIUNGAN ISLAND PIG Unidentified kind of pig, photographed on this island in the Philippines, at the end of the 20th Century.
scienceblog.com: tetrapod zoology: December 1st, 2010

BARBADOS SEA MONSTER A creature of bizarre appearance, seen off Barbados. It was said to be 40' long, its head was flat and its mouth was open most of the time.
P 56: 11

BARREN COUNTY SNAKE A giant snake has been reported from this part of Kentucky.
SKM: 118

BARRIO CORTES FIGURE Large humanoid creature with green eyes, seen by at least two witnesses in Puerto Rico in January, 2008.
www.iraap.org

BASAN In Japanese legend, a fowl of the chicken kind. It breathes fire, but happily this sort of fire does not burn.
Wikipedia

BASHE A monstrous snake in Chinese mythology. It was able to swallow an elephant.
Wikipedia

BEAMAN MONSTER   A creature supposedly having characteristics of hominid and wolf reported from the Kansas City region from the early 20th Century. There is a rumour that it is descended from an escaped gorilla.
www.unknowncreatures.com

BEANNACH-NIMHE   A beast in Scottish folklore. Its name means poisonous horned one.
GNB: 190

BEAR-LIKE CREATURES   In October, 2009, a motorist at Spring Lake Park (Minnesota) saw two figures looking like mutant bears beside the lake. One stood up. The witness saw they had human features with a snout. They had claws on their fingers.
ufoinfo/humanoid: 2009

BEAST OF ANKERDINE HILL   This peculiar animal, seen in England, was described as half fox-cub and half wild boar. Its nose was long and its skin was brown and mottled. It had a long tail.
GW: 70

BEAST OF BENAIS   An unidentified French beast whose depredations lasted from 1693-4. It flourished near the village and forest of Benais.
VIC

BEAST OF BLUEWATER   The unidentified tracks of an animal were found at a golf club at Gravesend (Kent) in 2009. The tracks are not inconsistent with those of a wolverine or fisher.
A&M 48: 47

BEAST OF BOAZ   A large creature with black fur reported from West Virginia. Standing on two legs, it is 7' tall. It has sharp teeth and a snout. It also has the power of human speech.
WT: 79-81

BEAST OF BONT   Unknown creature which killed many sheep in Wales in the 1990s.
www.paranormaldatabase.com

BEAST OF BOWMAN HILL   This creature was observed in Bucks County (Pa) in 1977 by one J. Lackey, who reported that its face resembled a monkey's, it had a body like a greyhound and was the size of a normal dog. Its tail and legs were long and it was the colour of a deer, from which one would infer it was a shade of brown.
Crypto Squad USA: October 6th, 2011

BEAST OF BRENTWOOD   An unidentified animal that killed rabbits and cats in Essex in 2009.
A&M 48: 50

BEAST OF BRYMBO   In 1985, two women at this Welsh village encountered a creature that resembled a cow. It was standing on its hind legs and appeared to be 6' tall or taller. One

Malcolm Jones, in the same vicinity, reported a big animal with shaggy hair in 1971. It stopped in the middle of the road, which it was crossing, and stared at him.
www.uncannyuk.com: May 4th, 2008.

BEAST OF CAEN A French beast which resembled a wolf but had some physiological differences, such as a red tail. It killed about thirty people.
VIC

BEAST OF CHARTRES An unidentified beast which killed a boy at Chartres in 1581.
VIC

BEAST OF CHIPLEY A beast whose tracks resembled a dog's was on the prowl in this part of Georgia (USA) in 1923. Despite the tracks, observers said it was much larger than a dog and had a different gait.
V: 208-9

BEAST OF HONG KONG An animal with grey hair which attacked gardener Law Chiu in 1955. There was another possible sighting by a woman, who said the creature proceeded on all fours.
www.naturalplane,blogspot.com: September, 2010

BEAST OF LYON This French animal had a career that stretched from 1754-6. It had some lupine characteristics, but many suggested it was an hyena.
VIC

BEAST OF VEYREAU Mystery wild beast recorded in the department of Aveyron, France, in 1799.
VIC

BECKERMET HUMANOID This creature with ginger hair was reported in Nursery Woods (Cumbria) in 1998 and 2005.
www.paranormaldatabase.com

BECKERMET PTEROSAUR A creature that looked like a pterosaur was reported over Nursery Woods (Cumbria) in 2006.
www.paranormaldatabase.com

BEDFORD FLYING SERPENT A huge serpent with glistening scales and forked tongue was reported by Lee Corder at Bedford (Iowa) in 1887. He at first mistook it for a buzzard.
HR: 230

BEIST MHOR ANAGNATHACH This was a beast in Scottish lore which sucked in its victims. If it came on a hapless ploughman, it would suck in man, plough and horses.
GNB: 189

BELIZE HUMANOID Belize is reputedly home to humanoids with red-brown hair.
AE: 43

BELL CITY ANIMAL An animal was reported at this Missouri location in 1958. It was striped, its tail was long and bushy and it was described as larger than a calf.
V: 350

BELLINGHAM BAY CREATURE An unknown creature observed in this bay in Washington state in 1895. It was estimated at 150' long and, additionally, with a neck, held upright, 30'-40'. It is important to remember that witnesses in these circumstances are likely to exaggerate measurements.
SM: 35

BENN The stone upon which this monster was supposedly killed was shown in Meath, Ireland. The creature had four heads and one hundred and forty legs.
GNB: 189

BENNINGTON MONSTER Near Glastenbury Mountain (Vermont) an unidentified monster with glowing eyes and of considerable size was said to exist. In the 19th Century it attacked and knocked over a stagecoach. Its footprints were unidentifiable. In 1934 a skeleton was found which was supposed to be the monster's, but turned out to be that of a cow.
VMG: 48

BERGEN COUNTY MONSTER In the northerly part of New Jersey, a girl answering her door saw this creature. It was about 3.5' in height, furry, standing on its hind legs with short forepaws. Its mouth seemed to bisect its face and it had sharp teeth. Its red eyes were swirling, hypnotic. The girl ran into the house. Her sister also had a sighting. The first mentioned witness, when an adult, claimed to have heard the sound of the creature outside.
MNJ: 70-72

BERKELEY SQUARE THING No 50, Berkeley Square, London, has a reputation for being haunted by a number of ghosts. A certain room there is rumoured to be downright dangerous. In the 1840s Sir Paul Warboys was egged on to spend the night there. The landlord, hearing a bell ring and a pistol shot, hastened to the room to find Warboys dead with a look of horror on his face. A story of which variants exist had two sailors staying in the room. One was killed, the other not. The Thing to which these actions have been attributed has been described as a collection of shadows. It has been described as shapeless, making a gruesome sloppy noise. One witness described it as looking like a small octopus. This has led to the suggestion that it is a cryptid, a small freshwater octopus that enters the building through the plumbing. Perhaps it has descendants, keeping the tradition alive.
www.americanmonsters.com: January 20th, 2010

BHOOTBILLI An animal whose name means 'ghost cat' and which has been allegedly seen near Pune, India. Its face is doglike, its back is mongoose-like, it has a long tail and it is of

plump configuration. It has been seizing pigeons and a goat.
www.cfz.org.uk: 11th November, 2010

BIDDRINA A reptilian monster, blue/green in colour, associated with the Italian province of Caltanissetta. Some say it is a hybrid of a dragon and a crocodile. Two specimens were allegedly killed in the 1950s by shepherds at Mount Saraceno.
Italian Wikipedia

BIG BLUE An unknown creature said to occupy the Big Blue Pond in Iowa. It may be a snapping turtle.

BIG ONE A large monster said to dwell in Lake Usmas in Lithuania. It may be an unknown species of turtle.
WW: 113

BIG SPRINGS LIZARD MAN A lizard man seen by a boy drinking from a pool at this locale in Texas. When the lizard man saw him, he entered the pool and endeavoured to swim towards him. The boy fled.
R: 62

BIRDMAN OF ALLENDE Near Allende (Nueva Leon), Mexico, on July 20th, 1994, a farm worker passing a graveyard encountered a creature which he described as half-man, half-bird. It walked past him, then spread its wings but did not take off, merely continuing its perambulations.
www.iraap.org: rosales

BIRDMAN OF MADISONVILLE This appears in fact to have been a large bird seen in this section of Tennessee in the 1960s and early 1970s. It seemed too big to be a bird. When on the ground, it appeared to walk like a man. Sightings had ceased by 1975.
Alternative Perceptions Magazine online: October 7th, 2007

BLACK BEAST OF SNAKE MOUNTAIN Snake Mountain (Vermont) was apparently home to a monster in the 1920s and 1930s. It was reported to have attacked a car by landing on its roof and to have chased a cycling child. Although several people saw this creature, descriptions appear to be lacking.
VMG: 45

BLACK BULL OF HAWK'S TOR A phantom bull in Cornish lore. If a certain loganstone at Hawk's Tor is touched at midnight, the Black Bull will appear.
fortean.wikidot.com

BLACK BULL OF MYLOR A phantasmal bull of Cornwall, said to bellow and have fire coming out of its nostrils.
fortean.wikidot.com

BLACK CREATURE A large animal seen off the Faroes in 1953.
WW: 44

BLACK FOSSA The fossa is a fierce creature (*Cryptoprocta ferox*) found in Madagascar. Though it looks like a cat, it isn't one. Black specimens of this creature have been reported, but not yet confirmed.
Shukernature

BLACK HUMANOID This 1m tall creature was observed in 1930 by Aida Semyenovna Sidorenko in her home in Moscow when she woke up. The humanoid seemed to have small wings. It flew off.
JHS: 1:3

BLACK MONKEY Cryptozoologist Richard Muirhead found a mention of someone in Hong Kong who kept one of these as a pet. What it was and whether it was a known Hong Kong species remain uncertain.
www.cfz.org.uk: 10th October, 2010

BLACK PARKS CREATURE Near Inverlochy Castle, Scotland, these animals, two in number, were seen by three wayfarers. Their bodies looked like greyhounds, but they were larger; they had sheep-like faces and no tails.
P 56: 11

BLACK PIG An animal of Irish legend, supposedly responsible for carving out the Black Pig's Dyke, a series of Iron Age earthworks to the south of Ulster. It was supposed to have been an enchanted schoolmaster.
MAI: 61

BLACK PUMA It is doubtful if black pumas exist in any considerable number and they are presumably not the mystery big cats seen in modern times. In 1843 a genuine black puma was killed in Brazil by a man named Thompson.   A genuine black puma was shot in Costa Rica in 1959 and photographed. A supposed black puma was seen in the Netherlands in 2005, but it turned out to be the offspring of a domestic cat and a wildcat.
MAL: 164

BLOOD-SUCKING SPIDER This fierce arachnid, which could grow to the size of a piglet, was believed to be found in Clare, Ireland. It would attack children.
MAI: 84

BLUE ALBATROSS This strangely coloured bird was observed in New Zealand in 2008. There have been other sightings (perhaps of the same bird). It is possibly a new species.
cryptodane.blogspot.com

BLUE BEE These insects were reported by someone who claimed to have seen them in Sydney, Australia, as a small child. They may, in fact, be a known species, but the witness

failed to identify them.
www.cryptozoology.com: July 10th, 2009

BLUE CREATURE   About 1999, two children in Australia reported seeing a blue creature that looked in some respects like a miniature lion, 6" tall and lacking a tail. The outer face was dark blue, the remainder of the animal lighter blue.
About.com

BLUE DEVIL   An animal of the dog family reported in Webster County (WV) from 1939-40. It had a bluish colour, a canine face and was the size of a pony. The killing of sheep in the area was ascribed to it. E. Corley thought an animal he had killed, which could not be certainly identified, was the creature.
MAWV: 81-82

BLUE DOG   An animal which appeared in Texas in 2004. There have also been reports of the creature from Alabama and Kansas. For some reason it has become identified as the chupacabras of Latin America, but this is a complete error. Because of their complete hairlessness, it can be inferred they are not suffering from mange, as mange that severe would render them immobile. Besides, a mother with young has been reported - it, too, was hairless. The possibility that photographs of dogs or coyotes with some hair or concealed hair suffer from mange cannot be ruled out.

The first known Blue Dog was shot by Devin McAnally in 2004 at Elmendorf, which lies to

the south of San Antonio. Dr Phyllis Canion of Cuero has a stuffed and mounted specimen, (opposite) from which one can judge that the animal is not a coyote. The creature in real life had blue eyes, which were replicated by the taxidermist. The skin was smooth, the torso looked like a deer's. Unlike a dog, it had slight shoulder blades and pelvis. However, DNA research has shown that neither of these specimens constitutes a new species.

As some of the animals have pads on the buttocks and some not: this may be a feature of sexual dimorphism.

DNA research has indicated the animal is definitely a canid. The Elmendorf Beast has been shown by genetic testing to have been a dog, while the Cuero creature was identified as being a cross between a Mexican wolf (*Canis lupus bayleyi*) and a coyote (*Canis latrans*). It is hard to account for a Mexican wolf's presence in this part of Texas. Some of the Blue Dogs may show, preserved in the genes of feral dogs, features of pre-Columbian hairless dogs.

The idea that some of these Blue Dogs are reptiles with canine features seems unlikely.
A&M 48: 65-77

BLUE EYES A creature without hair, with blue eyes and blue skin, seen in South Tyneside.
www.cfz.org.com: January 10th, 2011

BOAMAN A legendary creature of Maryland which seems to be unpleasant. It is half man, half snake.
marksofmessengers.blogspot.com

BOB-TAILED WOLF An unusual wolf, huge in size, with a black heavy coat and a bob-tail. It was shot in Boise National Forest (Idaho) in 1909. The remains, though allegedly sent to Washington, have disappeared.
FT 261: 25

BOIUNA A large black snake supposed to live in the Amazon. It is also known as *Cobra-Grande.*

BOKAAK This bird was once said to inhabit the Marshall Islands, but what it is is unknown. It is perhaps called by a different name these days.

BOMOSEEN EEL Lake Bomoseen (Vermont) is not only famous for the giant rabbits said to occupy an island in its midst, but also for a giant eel, 20' long, said to inhabit it.
VMG: 82

BOOGIE DOG A dog with a lion-head reported from Stroudsburg (Pennsylvania).
www.mysteriousbritain.com

BOOJUM A term applied to a specific animal, perhaps a Bigfoot, in North Carolina. This specimen was first reported in the Eagle Nest Hotel, which was built near the start of the 19th

Century. There it was given the name of *Boojum*, which had originally used by Lewis Carroll for a fictitious beast. The animal was described as 6'-8' tall, with grey hair. It was supposed to have married a human girl called Annie. Boojum used also to collect emeralds, which were easily found in the mountains.
www.northcarolinaghosts.com

BOOTED LYNX   An animal described by J.D. Bruce in *Travels to Discover the Source of the Nile* (1790). Bruce describes the animal as dirty grey on top, dirty white underneath. He suspected the animal was a scavenger. The creature may have been a specimen of the African wildcat (*Felis sylvestris lybica*).
www.cfz.org.uk: January 3rd, 2012.

BORLEY RECTORY INSECT   Borley Rectory was a house with a reputation for being haunted. Margaret Wilson, an artist, in 1938 saw this creature. It was 3" long with huge eyes. It was coloured black. She hit it, whereupon it struck the floor and disappeared.
www.cfz.org.uk: 28th September, 2010

BRAND-CATTLE   In the folklore of Sweden, the cattle of mermaids, which dwell under water.
NM: 282

BRAZILIAN CHUPACABRAS Here we have yet another creature classed as a chupacabras, even though it differs greatly from the usual description. This term seems to be becoming used for any unidentified animal in certain quarters.

Due to animal deaths a party of men holed themselves up in a shack at Capivari, Sao Paolo, to wait for the predator. During the night something started crashing against the shack. Looking through the window, they saw what appeared to be the culprit, covered in black hair, holding itself up against a tree. They estimated that it was 1.8m tall. It then rushed at the shack on all fours and barged against the door before making off into the woods.

Later the same men found a dead creature in the woods. It was hairless. They could not tell what it was. It had three bullet holes in it.

No further information about this story seems available. We seem to be dealing with two creatures here, or, if not, one with fast-acting alopecia.
www.iraap.com: rosales

BRAZILIAN GIANT W. Turner (died 1568) said that Brazil harboured a race of giants, the backs of their heads being flat. The women had black hair which was long and coarse.
PM

BRISTOL FISH This curious creature was reportedly caught in 1607. It was endowed with hands and feet. It was 5' long and no one seems to have been able to identify it.
MSB: 167

BRITISH BEAR There appears to be a rumour that there is a colony of bears, keeping a low profile, in Exmoor National Park in Somerset.
MSB: 154

BRITISH CAPYBARA It is now possible that the capybara (*Hydrochoerus hydrochaeris*), the world's largest rodent, has established itself in Worcestershire, according to reports. There was also a number of reports from Barnet, but some commentators suggested that they might refer to munjac, a breed of introduced deer.
GW: 76; *MAL:* 119-122

BRITISH COATI The coati or coatimundi now seems to have established a breeding population in Britain in the county of Cumbria in the north of England.
cryptozoology.com: January 8th, 2012

BRITISH LEOPARD A leopard and cubs were reported in Cinderford (Glos) in 2007. If leopards are breeding in the area, it could mean there will be a viable population in due course. A black panther (which is actually a black leopard) and a puma have also been reported from Cinderford.

Leopards (black or spotted) seem to be raising cubs in caves along the disused railway lines

near Stroud (Glos). A leopard hair was discovered in Devon at the Centre for Fortean Zoology's Weird Weekend in 2010.
GW: 48; *Mail Online:* November 25th, 2011

BRITISH LYNX There is considerable doubt about when the lynx died out in Britain, though the latest date is given about the 10th Century. However, the question has been raised regarding whether it died out at all or whether small populations survived until the present day. There have been numerous reports of sightings in Britain, e.g., at Burgess Hill in December, 2010, and Chichester in June, 2011. It has been suggested that lynxes have been deliberately introduced for hunting. However, now that hunting with dogs is illegal in England and Wales, such practices may cease.
sussexbigcats.blogspot.com; www.uksafari.com

BRITISH REDBACK The redback spider (*Latrodectus hasselti*) which is of Australian provenance, has been turning up increasingly in Britain due to stowaways in travellers' luggage. Redbacks are poisonous and dangerous.

BROOKLYN FLYING MAN This was reported by one William H. Smith in 1877.
MNJ: 25-26

BUCKSHAW BEAST A strange beast has been reported from the village of Buckshaw, near Leyland (Lancashire). It supposedly shows characteristics of hyena, wolf, wild boar and big cat. It has been blamed for the deaths of animals, including deer. An indistinct photograph has been taken.
Mail Online: January 20th, 2010

BUENA VISTA CREATURE This animal is so odd that there is a temptation to dismiss it as a hoax from the start. Its face was almost human and it had a moustache. Its size was that of a greyhound and it had a long, smooth tail. It was of a dull colour with black spots.

Near Buena Vista (California) it startled a horse drawing an equipage. It then departed and, as the driver of the rig made a hasty departure also, it stood on its hind legs. Others also claimed to have seen the beast.
V: 181

BUG-EYED MONSTER The *Kompira Maru* saw this creature in the seas of New Zealand. It in some ways resembled a crocodile, but had fins.
CFZY 2012: 174

BUGGANE A monster in Manx lore. According to legend, it tore off the roof of St Trinian's Church, annoyed by the bells.
paranormaldatabase.com

BUGGERMAN A creature believed to exist in Maryland. It is sometimes described as a black man covered in hair, sometimes as a ghost.
masksofmesingw.blogspot.com: January 2nd, 2010

BUIN MONSTER Chilean cryptid. A driver named Juan Barrios ran into this creature on 5th January, 2004. It had jumped in front of his microbus. It had a muzzle longer than a wolf's and a small hump on the back of its neck. It had black hairs. He procured a handful of the creature's hair which had been caught in the windscreen wipers.
www.bookofthoth.com/article *289;* www.rense.com/general48/hair [*]

BUKAVAC In Serbian legend, a water-dwelling six-legged animal with horns.
Wikipedia

BUKIT TIMAH MONKEY MAN A primate said to live in Singapore. Although reported from 1805, such reports are thin on the ground and may be inaccurate accounts of monkeys. It is supposed to be nocturnal, to walk upright and to have a monkey-like face.
Wikipedia

BULLOCH COUNTY ANIMAL An animal with the habit of chewing off the heads of dogs and humans which throve in this county of Georgia (USA) in 1919, but remained unidentified. It was shot at and not heard of again for a while, most people hoping it was now dead. However, it may not have been, as another unidentified beast was seen later.
V: 206

BUNGISINGIS (*also called* **Mahentoy**) A legendary tusked giant with a single eye in Filipino folklore. It has a tendency to laugh.
Wikipedia

---

* EDITOR'S NOTE: Both of these websites have since been taken down

BURRA HAAF MONSTER This Scottish marine creature was said to be 30' in length by witness P.F. Jameson, who espied it in 1903. It was also seen by J.M. Robertson and J.R. Anderson. It had an appendage that looked like a flipper on its head. It helped itself to the fish in the nets of the vessel *Adaloy*. The crew on one occasion had to push it away with a boat-hook.
NI: 123-5

C

CADDOS BIRDES Green horse in the folklore of Sardinia.
Wikipedia

CAGUAS CREATURE The subject of a vague report from Puerto Rico. It was a strange creature with a long neck, the witness said to have been a man named Gonzalez.
www.iraap.org: Rosales

CAI-CAI The Mapuche Indians of Chile say this is a sea-monster, part snake and part horse, sporting a mane. It tried to destroy mankind in a flood, but was prevented by a benign monster called Ten-ten. It is possible the cai-cai is a composite of two monsters.
PM

CALAVERAS ANIMAL An animal killed in Calaveras County (California) which could not be identified. Its head was foxlike, its tail was like that of a monkey and it had webbed feet.
V: 157

CALIMAYO Water-dwelling horses, quite widely believed to exist in South America.

CAMERON LAKE MONSTER Cameron Lake on Vancouver Island is, according to legend, bottomless, but it is in fact only 240' deep. It is said to contain underwater tunnels leading from it to Horne Lake. It is rumoured to contain a monster, a sighting of which, a long black shape, is said to have taken place in 2004. In 2007 witness C. Horvath thought she saw three objects, possibly animals, in the water. There have been other reported sightings.

A search for the creature using a fish finder scored two hits at Angel Rock in 2009 and another two hits were scored in 2010. As the lake is not very big, there is some possibility that the creature is a sturgeon, which was deliberately introduced to it.
www.cryptomundo.com: September 10th, 2010

CAMERON MOUNTAINS HOMINID A kind of man-beast reported from the South Island of New Zealand. Its existence has been inferred from footprints. Very little is known of this supposed creature.
nzcryptozoology.ucoz.com; CFZY 2008: 166

CANADIAN MOTHMAN This was seen in 2007 in Ontario by a witness (sex not specified) and neighbours. The creature was initially seen in tall grasses. It had red eyes, was about 7' tall, with long arms and a wingspan of about 8'. The witness and nephews fled, but during the night the witness heard scratching at the window. The creature was there. With a screech it departed.
About.com

CANNOCK CHASE FLYING MAN This being was sighted on February 8th, 2009, at Gentleshaw Common. There were five witnesses. Cannock Chase is well known for its strange phenomena.
ufoinfo.com/humanoid: 2009

CANNOCK CHASE MONSTER Cannock Chase (Staffordshire) which is pictured above courtesy Wikimedia Commons and Andy Stephenson, seems to be the focus for a great deal of paranormal phenomena. There was even said to have been a UFO crash landing there in 1964. In September, 2008, a local resident was chased in the early hours of the morning by some gigantic figure, perhaps proceeding by leaps, perhaps even by flying.
www.sundaymercury.net: November 14th, 2008

CANNOCK CHASE SNAKE This was seen in March, 2006, by ramblers, who said it was 14' in length and stood out from the local vegetation, which may indicate it was not native to the region, as it was not readily camouflaged.
TSW: 41

CANTON ZOO ANIMAL According to the *Straits Times* (7th January, 1936), an extraordinary animal was put on display in the Canton Municipal Zoo, China. It had apparently been caught in the mountains of the Tai Shan district. It had a cat's head, a horse's body and weighed 100 lbs. (Canton is nowadays spelled *Guangzhou*).

CAPACAUN Monstrous humanoid of Romanian legend, perhaps originally depicted with a dog's head.
Wikipedia

CARCANCHO A Wildman in the beliefs of the Mapuche Indians of South America. Although carcanchos are thought of as a species, individuals are supposed to lead a solitary existence.
PM

CARLOW ANIMAL This creature was reported in Ireland in 1924. It is said it looked like half a sheep and half a deer and shared a field with a donkey.
MAI: 83

CAROLINA BEACH CREATURE When near the Cape Fear River in North Carolina in April, 2008, the witness was surprised by a scream. He found it came from a 4' tall humanoid with leathery skin. The humanoid dived into the water.
www.iraaap.org: rosales

CARRAGUAR A large black felid, very fierce, reported from Colima, Mexico, in the 19th Century. This was its Indian name. Whites called it *renegron*.

CASAR HUMANOID A strange creature, generally of human shape but 10' in height, with a human face and yellow beard and covered in yellowish hair. It was seen by Tim Peeler of Casar (NC) in June, 2010.
JHS: 1:4

CASCADE MOUNTAIN WOLF This North American animal (*Canis lupus fuscus*) is supposed to have been extinct since 1940. However, a sighting of what strangely resembled one took place outside Concrete (Washington) in 2007.
www.cryptozoology.com: May 29th, 2007

CASSENYIE BEAR Legendary American bear.

CASTLE RING HUMANOID Castle Ring, a prehistoric hill-fort, is thought to have been occupied around 50 AD, perhaps by the tribe of the Cornovii. It is in the village of Cannock Wood, itself part of Cannock Chase (Staffordshire) where many strange phenomena have been reported recently. Alec Williams, driving past, saw a seven foot hair-covered humanoid cross the road into trees at the other side. The incident occurred on 9th October, 2010. Cannock Wood itself is regarded as an Area of Outstanding Natural Beauty.
www.cfz.org.uk: 13th October, 2010

CATALAN BIG BIRD This creature, black or grey in colouration, with membranous wings like a pterosaur, was reported over Catalonia for a period of about three months. There may have been several of them. There were hundreds of reports, with sizes of wingspan given from 3' to 15'. The first report came from Barcelona in 1990. The first mention was in the newspaper *La Vanguardia* which produced a flood of letters. At least one person called it a mutant pigeon.

CATALAN CHUPACABRAS Another animal identified with the chupacabras. It is a relief carved in Las Gavarres (Gerona), Spain.
criptozoologos.blogspot.com: April 3rd, 2011

CAVE DACHSHUND Small dogs said to live in Polish caves and to annoy potholers. Whether they are actually feral dachshunds or other creatures is difficult to determine from descriptions.
BR: 17

CENOCROCA A legendary creature made up of parts of many animals, boasting a solid bone instead of teeth. It contained elements of donkey, ibex and lion and had cloven hooves.

Italian Wikipedia

CHANNEL SEA SERPENT   Creature encountered in the English Channel in 1917. The lookout on the *Paramount*, an armed ship, sighted a huge creature like a giant conger eel, which reared out of the water on the port side. He estimated its length at 55', it had a spiny fin on its back and its colour was a dark olive green. The ship opened fire on the creature and it sank leaving blood about on the surface. This may be the same creature reported in 1957, a beast with a scar. It may also be identical with the Martello Monster.
www.cfz.org.uk: December 27th, 2010

CHEATHAM COUNTY CREATURE   A woman in this locale found a creature in her garage in February, 2010. It had long arms and short back legs and was quadrupedal. It had large eyes. Its colour was described as greyish.
JHS: l:1

CHEN   Highly poisonous bird of Chinese lore.
CMC: 248

CHICHIVILU   In the legends of Chiloe (Chile), a creature with a pig's head and a snake's body. It inhabits swamps.
LM

CHIH   A kind of fish mentioned in the Chinese *Shan Hai Jing.* What kind of fish it is is unknown.
CMS: 220

CHILEAN HIPPOPOTAMUS   A creature reported in the 18th Century. Unlike the hippopotamus of Africa, however, it had soft hair and palmate feet.
PM

CHINESE CHUPACABRAS   A strange brown-haired animal, compared with the chupacabras of Puerto Rico and which had been attacking chickens, was caught in a steel net in Suining, a Chinese village, on 24th March, 2010. Its length was about 60m, its tail 20m. It has been described as looking like a cross between a kangaroo and a dog.
cryptozoology.com

CHINESE PADDLEFISH   A huge fish (*Psephurus gladius*) of the Yangtze River. It has also been termed the elephant fish and the Chinese swordfish. The numbers were depleted by excessive fishing and, when the Gezhoubu Dam was built, this cut them off from their spawning ground. As a result, no young have been seen since 1995 and the last adult was seen

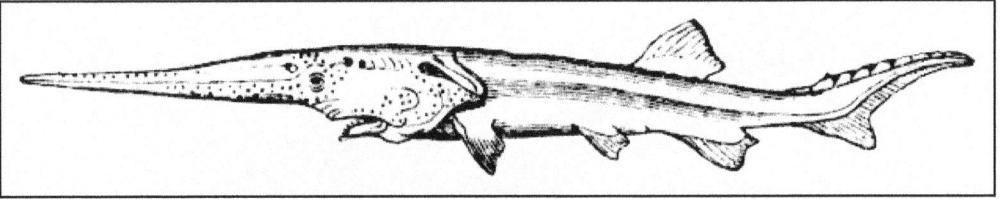

in 2003. This means it may now be extinct.
National Geographic News: July 26th, 2007; *IUCN Red List*

CHIPLEY ANIMAL A mystery animal in the vicinity of Chipley (Georgia) in 1923. It preyed on dogs and those who disturbed it said it was much larger than a dog. Its paw print was not unlike a dog's, but larger. It may have been fatally wounded by a farmer named Cands, who fired on it.
V: 208-9

CH'IUNG CH'I An animal resembling a winged tiger in Chinese lore. Humans figure in its diet.
CMS: 217

CH'O A creature which looks like a green rabbit with hooves like those of a deer in Chinese lore.
CMS: 229

CHOWA-CHOWA Small humanoids in Solomon Islands lore. They average about 4' in height. They have been reported on Guadalcanal and Makira.
SIM: 445

CHRYSOPTERA GIANT A giant enfolded in a large white cloud was reported in a garden in northern Greece in 1989. There was a bang and it disappeared. Footprints were left behind.
JHS: 1:5

CHRISTMAS ISLAND SHREW This animal (*Crocidura trichua*) is one of those which is possibly, but not certainly, extinct.

CHUHAISTER A kind of giant in Slavic folklore, reputed to live in the woods.

CLAY COUNTY CREATURE Clay County (Georgia) is supposedly home to a flying creature. It sits in trees, gazing at you with its red eyes. It is said to be about 4' tall.
www.answerbag.com: May 18th, 2008

CLEADON BHM A hairy biped reportedly seen near Cleadon Village (England) in 2003.
www.iraap.*org:* hallowell

CLENEDIN BIRD Large unidentified bird seen near Clendenin (West Virginia) in 2007.

The witness said it resembled a picture he subsequently saw of a teratorn (depicted above by Nobu Tamura). The plumage was dark brown or black, the beak black, the head bald and the neck encircled with yellowish feathers. Teratorns are supposed to have died out in the New World five million years ago.
MAWV: 17-18

CLIFTONVILLE SEA MONSTER A short sighting of this creature off the Kent coast in 1950 does not seem to have led to a description.
www.paranormaldatabase.com

COBRA ENCANTADA A beauteous woman turned by enchantment into a snake in the folklore of Brazil. She guards a treasure. You must break the spell to obtain the treasure and marry its guardian.
Wikipedia

COCOS ISLAND WILDCAT In 1906 a party of treasure hunters who had been on the Cocos Islands produced a wildcat they had found there and domesticated. The Cocos Islands are not supposed to harbour wildcats. What it actually was is unknown.
www.cfz.org.uk: 14th November, 2011.

COLD HESELDON HUMANOID At this location in County Durham, England, in 2009 a couple walking in a lane observed a humanoid with glowing red eyes. It screamed and absquatulated.
ufoinfo/humanoid: 2009

COLLET LIZARD A lizard with wings allegedly seen by certain schoolboys near a deep pond called the Collet at some stage before the American War of Independence. The location was in New York state.
HR: 243

COLO COLO This animal, in the folklore of Chilean Indians, looks like a long feathered rat or a long mouse with a head like that of a cock.
Wikipedia

CONANT ANIMAL A creature seen at Conant (Ohio). It was about 18" tall and 3' long, with a tail 2' long. It was dark red with yellowish stripes.
V: 496-7

CONDON FLYING HUMANOID This was reported by two witnesses in Queensland in April, 2008.
www.iraap.org: rosales

COOPERSTOWN GIANT In 2007 a driver who lived in New York state at about 11 p.m. saw a giant that resembled the Cardiff Giant (a famous hoax) striding through the woods.
A&M 48: 44

COPPERY THORNTAIL This bird (*Discosura letitiae*) may still exist It is only known from two old specimens from Bolivia, and is illustrated here by John Gould.

CORKYFIRE On the Isle of Skye, it is believed the legendary water-bull of the sea will sometimes mate with domestic cows. The hybrid they produce is given this name.
GNB: 74

CORNING FLYING SERPENT One Ford Ewalt, a man of sober habit, together with his wife, saw a flying snake at this Iowa location in 1926.
HR: 246

COW BEAR A legendary bear of the United States, said to be large and dark. It has a taste

for cow or moose. There is a patch of light hair on its throat.
V: 120

COW-CREATURE Such a creature, of bovine aspect but bipedal, was reported in Maryland in 1978.
marksofmessengers.blogspot.com

COXE'S CREATURE The angler J.A. Coxe in his book *Men, Fish and Tackle* (1936) reports a strange creature off the California coast. It had a reptilian head with coarse, reddish bristles, standing at least 10' out of he water. Its eyes were the size of dinner plates.
Coxe's book has been republished by Coachwhip Publications

CRAIGSMERE BEAST Unknown animal observed off the Florida coast by a ship called the *Craigsmere* in 1920. Its head was at a distance from its body. It had dorsal fins and was partially under water.
FUW: 24-25

CREAMY WHITE BEAR An unknown kind of bear, relatively small, was to be found in Montana in the 19th Century, where at least three were killed.
V: 356

CRITO A screaming creature, heard but never seen, in Honduran folklore.

CROCODILE-FROG An unidentified beast known to the Seluks of Sabah and at least once observed by a European. He could not discern the head, which was buried in a wild pig it was devouring, but native information tells us the head is like that of a crocodile. The body has black spots and brown scales and the hind legs are long. There is no tail.
karlshuker.blogspot.com: May 31st, 2011

CUELEBRE A kind of dragon in Spanish legend. They are believed in in Asturias and Cantabria. Each keeps a xana, a beautiful fairy, as a prisoner. Cuelebres eat men. They do not die, but grow old and eventually have to fly off to a draconic Elysium.
Wikipedia

CUMBERLAND DRAGON *see* **Goosefoot**

CURRANE DUV A sea-monster in the belief of the inhabitants of Kerry, Ireland. It is described as having a mane. It is supposed to be 15' in length.
MAI: 56

D

DAISY HILL CAT A very large felid which attacked a cow in Victoria (Australia). Its skin was dark grey with some ginger. The witness felt the cat was some kind of hybrid.
ABC: 300-322

DANCING CREATURE A driver at night in West Virginia, who had become lost, encountered a short animal with long thick white hair. It was jumping in the air, spinning and landing. It then made off into the bush. A minute or so later it emerged and attacked the driver, its hair standing out. The driver jumped into the car, at which the creature threw gravel. The driver then sounded his horn. The creature fell over, shouted and ran off on all fours.
WT: 75-76

DEATH VALLEY CREATURE In April, 2009, four teenagers were in a vehicle in this California locale which seemed to be affected by some sort of magnetic force. Subsequently they saw a strange humanoid looking through the window. The face seemed featureless, but its head gave off a faint bluish light. One of the teenagers opened fire and it clutched its head with long, thinnish hands. As it ran off, they saw its arms and legs were very long. They estimated its height at 7'5".
ufoinfo.com/humanoid: 2009

DEEP CREEK CREATURE A possibly reptilian creature which may have been able to fly reported from this California locale.
FZ: January 21st, 2012

DERBYSHIRE HUMANOID This creature was seen from the side of the A57. It was 2.5m tall and had long brown hair. This happened in 1991.
www.paranormaldatabase.com

DEVIL BIRD A creature in the lore of the Indians of British Columbia. It is supposed to have a wingspan of 30'-40', to capture people (apparently to eat), to hunt in open areas by night and to inhabit caves. The Indians believe it is sent from Hell each night. Its diet includes moose and wapiti. Its tracks are distinctive and have been noted in the snow. It can sound like a dog barking, a woman screaming or a baby crying.

G. McIsaac feels it is a pterodactyl. McIsaac tells of a woman he knew who had to take shelter in a dog house to avoid capture by one of these creatures. A friend of his saw one walking on a bridge over the Findly River. He also tells of a traveller who was trapped in a car all night by a devil bird.

We cannot rule out the possibility of its being a surviving form of pterodactyl or some other kind of pterosaur.
BH: 1-40

DEVIL-LIZARD   These creatures have been reported from northern Argentina.
FZ: July 15th, 2011

DEVON NESTERS   The writer claims he saw a large nest resembling a wasp's near a Devon river (unspecified) spanned by a wooden bridge. The nest fell onto the bridge as a result of sticks' being thrown at it. The nest grew inside, pulsating the while, when a dog pushed it into the river. Out came horrific inch-long creatures resembling snakes with large red eyes and spined backs. They flew off. The writer is sure they were the offspring of some cryptid, but it is a pity he furnishes no details of the time and place of this incident.
About.com: paranormal phenomena: April, 2006

DIAMOND ANIMAL   In 1952 a creature that was black with yellow striped underparts was seen in the Diamond area of Missouri.
V: 347

DIAMOND-SNAKE   A term used in the Solomon Islands for a kind of unknown python which has a stone resembling a diamond on its head. The term is also applied to UFOs.
SIM: 143

DIXONVILLE CREATURE   A newspaper report claimed that in 1944 an artificial tunnel was discovered in the vicinity of this Pennsylvania town. In it was found a creature that looked not of this world.
News Extra: July 14th, 1974

DOE LAKE CREATURE   An unidentified creature seen by anglers in a boat in Middle Doe Lake (Ontario) in 2010. It looked like a big rock with protruding smaller rocks. When the witnesses approached, it submerged.
Psican

DOG BEAR Legendary American bear with a pointed snout and long legs.
V: 120

DOG-LIKE ANIMAL (MANITOBA)   A shaggy brown animal, reported from Manitoba about the year 2000.
V: 312

DOG-LIKE ANIMAL (TEXAS) A strange looking animal with doglike features, but with a face in someway resembling a deer, has been spotted the east of Texas in 2010. A number of photographs have been taken. No firm identification has been made.
americanmonsters.com: July, 2010

DOG-LIKE CREATURE This was seen in 1978. It exceeded a fox in size. Its tail was long and broad. It attacked and injured a pony at Prestonburg (Kentucky).
SKM: 88

DONEGAL WINGED MAN   Donegal (Pa) was the scene of an encounter with a winged humanoid who flew over a road.
SI: 140

DONES D'AGUA   Water creatures, looking like beautiful women, sometimes half fish or half bird.
Encyclopedia Mythica

DORRINGTON HUMANOID   This creature of Lincolnshire, a sort of man-beast, was reported to the police in November, 1970.
www.paranormaldatabase.com

DORSET WILDMEN   In days agone (unspecified), large hairy wildmen were said to live on Yellowham Hill in Dorset. They were reputed to abduct women.
www.paranormaldatabase.com

DOS PALMS SERPENT   A huge serpent about 30' in length with small wings near its head and its tail dragging on the ground was reported by the crew of a train travelling through this part of California in 1882. The train removed part of the animal's tail and the serpent pursued the locomotive, some of those aboard discharging firearms at it.
HR: 233

DOUGLAS COUNTY CREATURE   Unidentified animal discovered as road kill in Douglas County (Minnesota). It was white and one witness said it had an almost human look. It has also been suggested it was a governmental experiment. Scientists are allegedly nonplussed regarding its identity.
DiscoveryNews: August 8th, 2011

DOWNS ROAD MONSTER   This creature is described as looking like a Bigfoot, but only 4'-5' tall. Downs Road is in Bethany (Connecticut). Various strange phenomena have been reported in the area.
www.orangepatch.com: February 24th, 2012

DOX   Dog-fox hybrid. The evidence for such a creature's ever having existed is flimsy. In 1907 a possible one was reported as being killed in Warwickshire. The tail was thick and white. Amongst British gamekeepers there is a folklore belief that dog-foxes can breed with terrier bitches.
FS: 49

DRAGON-SNAKE   This creature is believed in by islanders on Guadalcanal. M. Boirayon identifies sightings as those of UFOs.
SIM: cap. 2

DRAGUA   The Albanian dragon, having wings, legs and a horn. Its lifespan is a century.

DRAKE'S POOL MONSTER A large but ill-described creature was supposed to have been in this pool at Carrigaline (Cork), Ireland, in the 1860s. It would come on land and hunt for food.
www.paranormaldatabase.com

DRUK The Thunder Dragon, believed in by the people of Bhutan.

DRY GULCH CHUPACABRA A name given by locals to a strange animal captured near Adair (Oklahoma) in 2010. Though it is said to be a mangy raccoon, not everyone is sure about this.
NewsOn6.com: March 2nd, 2010

DUBBLEDAM GHOST BEAR In the 19th Century a wayfarer noticed a small dog behind him in the village of Dubbledam in the Netherlands. It turned into a bear. Happily he had an iron implement upon him and the bear left him alone.
www.cfz.org.uk: November 25th, 2010

DUNKETTLE GIANT EEL This was supposed to have been a huge creature that ate sheep and cattle in Co Cork, Ireland. Its ultimate fate is unknown.
www.paranormaldatabase.com

DUNROBIN SEA-SERPENT This beast was seen off the Scottish coast a number of times about 1873.
www.paranormaldatabase.com

DURHAM CREATURE Seen in England in the vicinity of Cold Heselton in 2009, this humanoid was 7' tall and black in colour. The sighting occurred in 2009. There were two witnesses.
HSR: 2009

DUTCHMAN CREEK ANIMAL A mysterious animal seen in Nelson County (Kentucky) in the mid-19th Century. It was witnessed by H. Smith. It was very large, spotted, seemed canine and had no visible head or tail.
www.kentuckybigfoot.com

DZAINOSGOWA In Seneca lore, a gigantic blue lizard, ultimately killed by a meteor.
MAP: 19

E

EAST RIVER CREATURE   A strange animal found dead on the banks of the East River (New York) in 2012. Experts say they are baffled by it. Though it looks somewhat porcine, it has toes rather than trotters. It turned out (allegedly) to be a dog.
wild mysteries.blogspot.co.uk: July 30th, 2012; www.telegraph.co.uk: August 7th, 2012

EASTERBEAST   This was the name given by B. Calvi and his wife to a curious animal that ran in front of their car. Its head was like a dog's, its ears like a fox's, its nose was pointed and, although its front legs were long, its back legs were short. Its eyes glowed fluorescent orange. The witnesses were inclined to discount a coydog identity. Some weeks later, two of the animals were seen by two other witnesses.
cryptomundo.com: August 16th, 2010

ECIJA ANIMAL   This animal was reported near the banks of the River Genii near Seville, Spain, in 1955. It was doglike, resembling a wolf dog (*perro lobo*) but with a more elongated body. Its main colour seems to have been blond/white. Its head was thin and snouted, while it had two horns. Its reaction to gunshots was to jump into the river (very sensible, if you ask me).
Criptozoologia en España: May 25th, 2012

EDEN (UTAH) FLYING SERPENT   A 60' serpent flew in what seems a rather sedate fashion over this town in 1894.
HR: 238

EDINBURGH HUMANOIDS   On February 16th, 2009, the witness woke up to find two humanoids in his dwelling. They seemed able to converse and one also seems to have tried to talk to the witness.
ufoinfo.com/humanoid: 2009

EKEK   Winged humans in Filipino legend.

EL CARRIZAL APE   At the village of El Carrizal, Grand Canary, in 2003 a local witness noticed at night creatures resembling apes hanging from trees. They appeared to have no faces. There seems to have been a "missing time" element in the experience and everything went silent when the witness observed the creatures, notwithstanding the proximity of an airport.
cryptozoologos.blogspot.com.es: May 18th, 2012

ELEPHANT MAN   Creature reported from Sydney, Australia, in 1960. It was bipedal, its front legs being short. Its head was like an anteater's. It had a stiff trunk.
CFZY 2012: 215

ELGIN MONKEY-MAN   A creature reported in 2005 in Elgin (Illinois). Some speculated that it was a chimpanzee.
www.cfz.org.uk

EPPING FOREST CREATURE   A strange creature, about 4' tall, humanoid, covered in hair, seen in Epping Forest, England, in 2008.
www.iraap.org: rosales

ERIE CREATURE   A winged humanoid seen near Lake Erie. The wings were transparent, about 3" broad and rounded at the extremities.
JHS: 1:4

ETNA FLYING SERPENT   In 1887 a number of people in this Pennsylvania town are said to have seen a flying serpent about 25' long. It was black and, when it flew downwards, it would open its mouth, revealing a huge tongue.
HR: 235-6

EUROA BEAST   The reports of this beast come from Wylonemby Swamp, near Euroa (Victoria) Australia. The creature is described as being 30' in length with a head like a bulldog's. Reports featured in newspapers in 1890. The creature startled two young men by the noise it made. One of them, revisiting the swamp on the succeeding day, observed the bulldog-like head. A hunting party, bent on flushing out the creature, saw the tail, which they claimed was as thick as a man's thigh. Some seemed to have glimpsed the animal, saying it

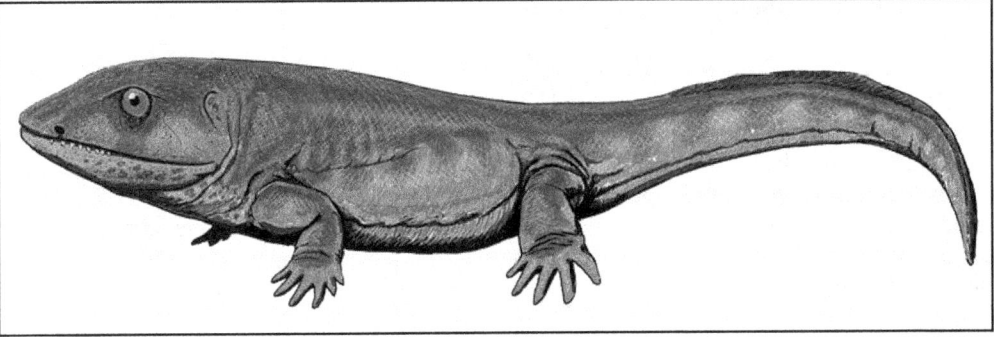

was brown on top while its underparts were yellow. There has been speculation that the animal was either a serpent or the supposedly long extinct prehistoric amphibian *Pederpes finneyae*. (See above) [*]
www.americanmonsters.com: January 4th, 2011

EVREN The dragon of Turkish legend, a large snake which shoots fire from its tail. These creatures are still believed to be found in the Altai and Tien Shan. In Turkic languages, the word *evren* can also be used for other reptiles.

F

FACELESS CREATURE A humanoid that seemed to lack a face. It had scaly skin and black hair. It was seen in New Jersey by a number of witnesses in 1966 and possibly earlier in 1965.
I: 255

FALCON CREATURE A silent creature seen gliding across the water at El Supi Beach, Falcon, Venezuela, in October 2008, at night. The creature was glowing.
www.iraap.org: rosales

FAROESE SEA-MONSTERS Sea-monsters have been reported to the north of the Faroes in 1869 and 1923. They resembled plesiosaurs.
WW: 44

FEI-YI In the Chinese *Shan Hai Jing* this name is applied to both a kind of snake and a kind of bird.
CMS: 248

---

[*] Long extinct is the right term. According to Wikipedia these creatures hail from the Lower Missippian, approximately 359 - 345 million years ago.

FENG-SHILT Gigantic boar of Chinese lore.
CMS: 203

FEOLAGAN An animal in which the Highlanders of Scotland believe, particularly on the Isle of Lewis. It looks like a large mouse. If it runs across a sheep's back, it will paralyse the poor animal completely, the remedy being if it then runs across the sheep's back in the opposite direction. Shepherds eventually are said to have evolved the technique of keeping a dead feolagan in a jar with salt. The feolagan would imbue the salt with some of its power, so if you found one of your sheep had been paralysed by another feolagan, you simply shook the salt over it to restore its mobility.
WI: 41

FERGUS FALLS CREATURE A dark brown, apparently bipedal, creature with protruding muzzle, seen at this Minnesota locale in 2009.
ufo.info/humanoid: 2009

FIJI CREATURE This animal was reported recently off the coast of Fiji and described as a 'sea-monster'. One witness said it had a strange, long neck and flippers. It was, however, only about 3' long, belying its monstrosity, unless it was the young of some greater animal.
P 51: 13

FINNISH HUMANOID This was 1m tall and seen by a woman who woke up, presumably in bed. It had very thin legs and seemed to generate some form of electricity. The incident took place near Varkaus in 2009.
ufoinfo,com/humanoid: 2009

FIOLLAN WORM This creature of the Highlands was reported in the 17th Century. It had a sharp white head, a red body and little feet. It was small enough to crawl under a human's skin, where it would travel around the body, causing pain and bruising. It would also lay eggs, leading to the victim having a whole population of the creatures beneath its skin. Some said it caused scrofula. However, the term *fiollan* was applied to a number of other creatures in Scottish Gaelic, including a worm and parasite generally.
WI: 47-8

FIRECO CREATURE An unidentified creature with yellowish-grey hair and a tendency to eat the brains of hogs was on the rampage in this West Virginia area in 1934.
V: 627-8

FIRE-SLUG This animal is said to look like a very large slug and to be found in Latvia. It leaves a trail of slime that chars everything it crosses by burning it away. The animal is also covered in this slime, so it should not be touched, unless you like burning your hand and reducing yourself to agony.
WW: 114-115

FISH-CREATURE This surprising creature jumped out of a hedge and dived over a cliff

into the sea. The witness, a shepherd, said it resembled a skate. The incident took place on the Atlantic coast of Ireland.
MAI: 65

FIVE FORKS ANIMAL This creature was seen in Georgia (USA) in 1896. Though described by such words as 'very frightful' and 'ferocious' we are not given any information regarding its actual appearance.
V: 207

FJORULALLI An Icelandic cryptid which lives on the coast and is thought to attack sheep.
AZ: 291

FLORIANOPOLIS CREATURE This encounter took place in the Brazilian state of Santa Catarina in 2009. A motorist saw in the dark a humanoid with an egg-shaped head, egg-shaped ears and long dangling arms. The creature was pale white.
ufoinfo.com/humanoid: 2009

FLORIDA DEER A deer with ears like a donkey, which defied identification and capture in Florida about 1997.
AZ: 38

FLYING HUMANOID In 1960 one of these was seen over Siracusa (Syracuse) in Sicily. In 1991 one with green hair and glowing eyes making robotic movements was seen in the Padua region.
CFZY 2009: 149

FLYING PINK JELLYFISH This was reported over Merton, London, in 2009. It was large and there was a pink haze around it.
PL: 58

FOGLIONGO A creature of Italian folklore, resembling the chupacabras. It is noted for killing poultry and exsanguinating them. It is particularly well known in Garafagnana (Lucca).
Italian Wikipedia

FOOPENGERKLE The name of this cryptid may have been humorous, but there is a tradition in Kansas that it occupies a sinkhole or drained lake, that it is serpentine and 15' long. It has been given the nickname of *Sinkhole Sam.*

FOX-FACED CREATURE An unknown creature with a fox-like face and a stubby tail captured at Au Sable Forks (NY) in 1943.
V: 398

FRANKLIN COUNTY CRYPTID On a property in Arkansas one Harley Edgin and a passenger saw two animals they could not identify, one larger than the other. They were the colour of a bobcat, with heads disproportionately big. The ears were about 6" in length,

topped by tufts of hair. The chests were massive, the front legs muscular. The back legs were not as long as the forelegs. The backs sloped downwards. Edgin claimed to have seen single animals on other occasions. Other witnesses have observed similar animals.
V: 148-50

FREDERICK REPTILIAN A humanoid reported in Frederick County, Maryland, in the 1880s. marksofmessenger.blogspot.com

FRESNO PTEROSAUR There were two of these seen in 1891. The bills were said to be long, the heads broad and the eyes large. Later witnesses claimed the bills were more like alligators' snouts.
LPA: 14

FUTA FILU A plump snake-like creature with stiff hairs. Its eyes resemble a cat's. It is believed in by the Mapuche Indians of South America.
PM

G

GABLE ISLAND MONSTER The Yagan of Tierra del Fuego believed in this, a gigantic man-eating seal or sea-lion. Local hero Umoara put an end to its career.
PM

GALLIPOLIS FLYING CREATURE In August, 2010, two women in a car in Gallipolis (Ohio) stopped the vehicle, as one (Kyra) had to purchase items from a store. Megan, her companion, thought she saw a large flying creature, perhaps a bird, though its skin resembled leather. They went to their hotel. During the night, Kyra heard scratching in the hall outside her door. Later she heard scratching outside her window and drew back the curtains to behold a bald humanoid with wings and bulging eyes. Then it spread its wings and took off.
JHS: 2:3

GATA CARRION In Italian folklore, a shaggy red cat which will steal the souls of children. It is supposed to be found in Bergamo and Cremona.
Italian Wikipedia

GATTO MAMMONE A magical cat of Italian folklore. Sometimes it is beneficial, but sometimes it disturbs herds of livestock.
Italian Wikipedia

GATO DE CHIFRES A horned cat, rumoured to exist in Indonesia.
Portuguese Wikipedia

GENIL CREATURE A mysterious animal reported from the Rio Genil, Spain, in 1955. It was said to be the size of a wolfhound with a long body, red back and white underside. It had a delicate head and a snout and two dark horns, about 8" long, protruding from its forehead. If by wolfhound either the Irish or Russian wolfhound is meant, both have fairly long backs already and attain considerable height. The idea was mooted that the beast was a manatee or coypu, but its description corresponds to neither of those. A further report stated it fed on peppers and poultry.
FT 261: 30

GEORGIA MYSTERY CAT An unidentified kind of cat with a bobcat's ears and a very long tail. Its height was 2.5-3', its length 4-5.5 feet. It was seen in 2004.
www.messybeast.com

GEORGIA SAURIAN In July, 2008, Y. Philips and his grandfather were hunting at night in woods in Georgia. To Philips' astonishment, they saw ahead of them what appeared to be a bipedal dinosaur, about 5' tall at the shoulder. It had a large claw on each foot and a stiff tail.
About.com: Paranormal

GERITS Monstrous creature resembling a tiger in Malay myth. There appears to have been only

a single one, which was trapped and killed by other animals.
PaM: 88

GERMAN MOTHMAN   An indistinct picture of something that looked like a mothman showed up on a photograph taken by Abbey Linfoot of York when she was visiting Nuremberg. She did not notice the object at the time she took the photograph.
www.americanmonsters.com: October 1st, 2010

GETZKO   The getzkos are wildmen in the folklore of Poland. They would walk both upright and on all fours. It is, curiously, believed that when they run, they do so in a sideways fashion. They are (or were) completely covered in fur, except for the eyes, noses, mouths and

palms. One wonders if the soles of their feet were uncovered, but unobserved. Their diet includes leaves and fish. They seem to have a rapport with other animals, but can show hostility to man. A sighting was reported in the 1890s at Orynka, now over the border in Belarus.
cryptozoology.com: *forums:* March 16th, 2006

GIANT ADDER These were supposedly to be found in Sussex in the 6th Century. Then, when a dragon was killed in the area, they disappeared.
K: 311

GIANT BEE This was observed in Australia in 1992. Its provenance is unknown.
www.cfz.org.uk: 30th September, 2010

GIANT BIRD
1. This huge black bird was seen by Blanca Trevino, as she was driving in San Antonio (Texas). It flew towards her windscreen and she estimated its head to be the size of a German shepherd's. (I refer to the dog rather than a human herder of sheep). Its beak reminded her of a parrot's. She had to swerve to avoid it.
MT: 12, 15
2 A huge bird with a head the size of a small dog reported over South Greensburg (Pa) in 2001. It flapped its wings slowly as it flew and sometimes glided.
RMP

GIANT CANADIAN INSECTS These were yellow and black, described in 1974 in Ontario.
www.cfz.org.uk: 30th September, 2010

GIANT CENTIPEDE Huge centipede rumoured to exist in the Amazon rainforest. Such a creature might have the potential to kill an adult human.
animal discovery.com/tv/lost-tapes

GIANT EARWIG An apparently extinct insect on the island of St Helena. However, there were some reported sightings in the 20th Century.
www.newanimal.org/giant-animal.htm

GIANT FISH A huge fish was seen off Tenby in the English Channel in 1902. Its length was estimated at 200'.
Western Times: June 2nd, 1902

GIANT FROG In Cherokee legend, a frog the size of a house. It was once seen at Frogtown (Ga), which was called in Cherokee *Walasi'ya* (frog place).
Mooney: 418

GIANT LOBSTER OF TROW ROCKS To the north-east of England in 1919, a large iron gate sank in the North Sea. The top of the wreckage was discernable and the story grew up that there was a giant lobster (or crab) beneath it. It seems that a student in 1960/61 reported some sort of monster which scuttled from the beach into the sea.

Although giant sea scorpions are supposed to have died out millions of years ago, if this is not the case, then this creature may have been one and taken up residence beneath the gate.
N&T: 223-229

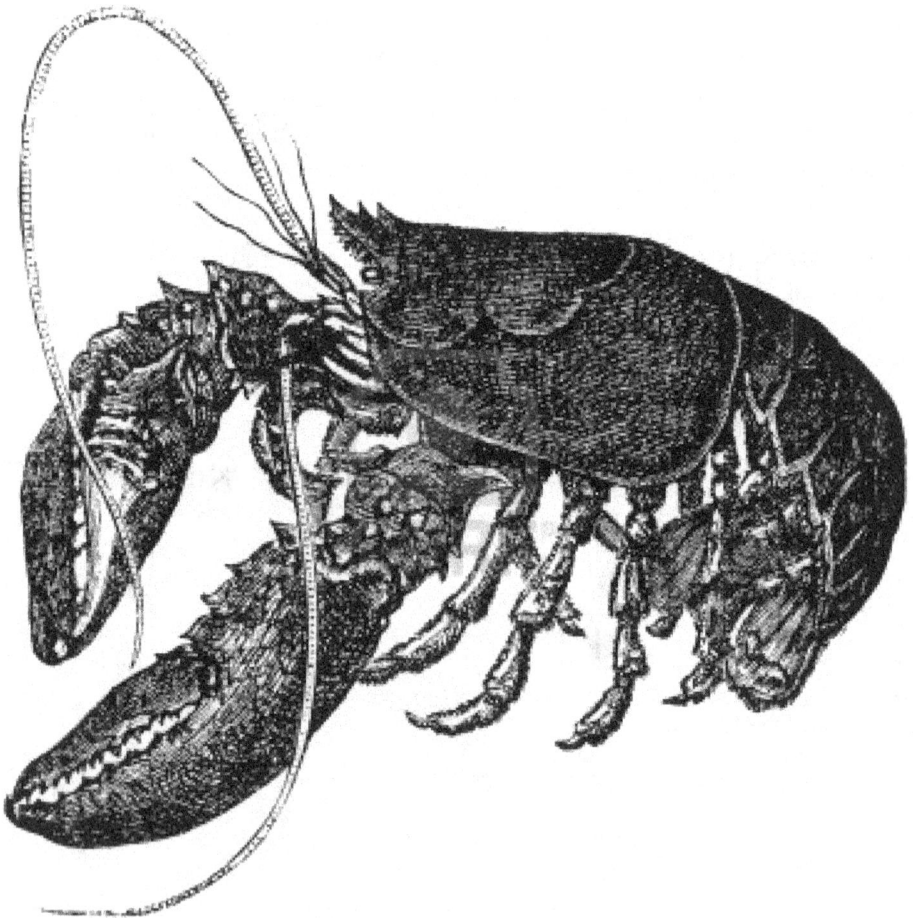

GIANT ORANGE INSECT   This unidentified insect, about a foot long, was seen in Alburg (Vermont) about 1980.
VMG: 55

GIANT PAPUAN SPIDER   This spider, jet-black and the size of a puppy, was allegedly seen during World War II.
FT 191: 54

GIANT PIG   A race of giant pigs was thought to have been found in Ireland. Their last refuge was Imokilly (Cork). The last specimen was killed and buried in a chest.
MAI: 89

GIANT SAWFISH   The ancient Romans claimed such creatures could be found in the Red Sea.

GIANT SEA-BIRD   The *Evening Post* in 1889 reported the attack of a huge bird on a teenage sailor who had fallen overboard from the *Talisman* off the coast of Chile. The bird, described as 'enormous' succeeded in taking off with the youngster and reaching a height of 30'-40'. A second bird also appeared. A sailor who had jumped in to rescue the boy was attacked by this. Captain Putt shot and wounded the first bird, which dropped its captive. A boat lowered from the ship was attacked by the second bird, which was at last driven off. All in the water were rescued. The first bird was captured with the aid of a lasso, but died three weeks later. It was left in Valparaiso for taxidermy.
JHS: 1

GIANT SEA-SNAIL   This beastie was reported from the Baltic in the 16th Century. Unlike most snails it had paws and the eyes were situated on either side of its head. It has not been reported in modern times, so French restauranteurs will not find it worth their while to embark on a hunting expedition.
Shukernature: April 16th, 2012

GIANT SEVERN EEL   Legend has it that this creature entered the Severn Estuary and ate so many fish that the fishermen drove it into Wookey Hole.   It is said it still occupies underground streams and cannot get out.
MSB: 167

GIANT SKUNK   A creature reported from Java, Indonesia, in 1977. These animals were said to equal a German shepherd (dog, not man) in size. One was allegedly killed. They are unlikely to be true skunks.
www.angelfire.com

GIANT VERMONT RABBIT   These animals are found on Rabbit Island in Lake Bomoseen, Rutland County, Vermont, and have also been reported elsewhere in the county. They have red eyes that glow.
VMG: 29

GIANT WORM These worms were reported in Ireland to Lady Gregory. One, nearly 8' in length, attacked a woman near Clough.
MAI: 100

GIGAT A larger version of the Gata Carrion. It seems to be regarded as a very fierce beast. It is found in the folklore of Lombardy and Sardinia.
Italian Wikipedia

GILYUK A gigantic humanoid in the legends of the Indians of Alaska. It was shaggy and ate people.
TG: 92

GIRONA GNOME Creature allegedly captured in 1989 in Spain, which died. It was 12cm long, bluish in colour, long-eared and it had a rodent-like snout. It may be an unknown animal, but was possibly also a deformed one.
Wikipedia

GISBORNE LIZARD On 28th September, 1898, the New Zealand Institute meeting was told of a large unidentified lizard seen near Gisborne.
nzcryptozoology.ucoz.com

GLAUCOUS MACAW This bird (*Androrhynchus glaucus*), which formerly was found in South America, may now be extinct. It subsisted largely on the nuts of the yatay palm, but many of these have been cut down.

GLOBE SWAMP MONSTER An unidentified animal, rumoured to exist in North Carolina in the early 20th Century. When one A. Britt killed an animal he could not identify but thought might be a jackal, he reckoned he had slain the creature. This occurred in 1914.
V: 450

GLOWING ANIMAL This glowing creature was seen at night by campers. It was the size of a large dog. On the campers' approach, it raced off. It jumped on a wall and stood on its hind legs. It was the size of a human. Unfortunately, I can find no clue regarding the date or location of this incident.
About.com: Paranormal Phenomena: *Camping Encounters with Monsters and Ghosts*

GLOWING HUMANOID Such a creature (or creatures) was (were) reported from Thompson (Connecticut) in 2009.
ufoinfo/humanoid: 2009

GLUHSCHWANZ A draconic beast with a glowing tail in German folklore.
German Wikipedia

GOEGOEH Alternative name for the orang pendek.

GOLDEN LIZARD   A lizard dwelling in a cave. It is said to chase cattle in Honduran folklore.

GOLDTOWN CREATURE   Strange howlings in this part of West Virginia in 2010 have sparked speculation that there is a mystery animal in the woods there. Some have opined it is a sasquatch.
cryptozoology.com

GOLLYWAMPUS A creature that looks like a big mink which is said to dwell in the water in the Ozarks.
V: 122

GONDWINISHIRE CAT   A mystery cat seen in Queensland in 2001. Although black, it had a white stomach. The witness was a man named Rush.
ABC: 343

GOON OF GUILLEMARD STREET   A creature with the head of a panther (=?puma) and the body of a gorilla with six arms and six legs which, according to rumour, was on the loose in Pensacola (Florida) in 1938.
www.cfz.org.uk: December 23rd, 2010

GOOSE LAKE MONSTER   This animal, reported in 1881, was between 15'-20' in length, hairy and had a head like a dog's and a tail like a catfish's. It was speared by a man named Osborne and, though it escaped, it was not reported again. Goose Lake is in Florida.
FUW: 51

GOOSEFOOT   A creature reported in 1794 in what is now Tennessee. It was bipedal, was covered with scales, its colours were black, brown and light yellow, it had a white tuft or crown on the top of its head and it left tracks like a goose. It was encountered by soldiers who discovered from Indians that the breath of this creature would kill a man unless he completely immersed himself in water once breathed on. This creature has also been referred to as the Cumberland Dragon.
www.strangeark.com

GOSH-GE   A near-invulnerable monster in the legends of the Aonikenk Indians of Argentina. It was protected by armadillo-like plates, rendering it almost invulnerable.
PM

GOSH-WHAT-IS-IT   This is the name a journalist bestowed on an unidentified creature cornered in El Cerrito (California) in 1955. It was brown, 16" tall, without a tail. It was not a skunk or raccoon.
V: 164

GRAFTON MONSTER   A monster, apparently headless but alive, white in colour, discerned by R. Cockerell, journalist, by the Tygart River at Grafton (West Virginia). The skin was not

unlike that of a seal. Others reported seeing the monster at other places down the Tygart.
WT: 42-9

GRAMPUS A creature supposed to have lived in a tree in the village of Highclere (Hants). It resembled a dolphin or porpoise. It was banished by a clergyman. Sir Stafford Northcote, Earl of Iddlesleigh (1818-1887) mentioned the grampus, which he called a *grumpus*, in his diary. To avoid confusion we should add that the term grampus has at times been used to mean the killer whale (*Orcinus orca*) and various species of porpoise. In zoology, there is a grampus genus, whose only member is Risso's dolphin (*Grampus griseous*).
www.americanmonsters.com: October 28th, 2010

GREAT BARRIER REEF MONSTER A sea-monster of prehistoric aspect reported off Australia by actor Ving Rames. The head was a rough oblong, the body like that of a catfish, he claimed.
torontosun.com: August 16th, 2010

GREAT LIZARD A huge lizard with a glistening throat and a penchant for sunbathing, that Cherokee legend once said lived on Joanna Bald Mountain (North Carolina).
Mooney: 408

GREEK HUMANOIDS Gigantic humanoids reported from the Peloponnese in 1978. The witness was driver A. Coulouris.
JHS: 1:5

GREEN APE An ape with green fur and green eyes which attacked a car in Elfers (Florida) in 1967.
I: 255

GREEN PHANTOM CAT This animal, which glowed, was reported at Balbriggan on the east coast of Ireland in 1996.

GREENSBURG FLYING CREATURE A huge creature seen near Westmoreland Mall, Greensburg (Pa). The sighting is undated. The beak was described as very large, evoking images of a prehistoric bird.
RMP: 100

GREENWICH CREATURE A strange creature shot in Greenwich (NJ) in 1925. The size of an Airedale, it had four webbed toes, yellow eyes and it hopped like a kangaroo.
MNJ: 69

GREY DHOLE A kind of wild dog reported from Burma, but unknown to science. There may be two varieties, one smaller than the other.
Wikipedia

GREY PANTHER Mystery felid reported at Onondaga Hill Road (NY) in 1949. It is not

unknown for a puma to be grey, but such specimens are rare - perhaps it was one of these.
V: 408

GRUNCH   Grunches resemble the chupacabra and are said to inhabit Grunch Lane in New Orleans.
Wikipedia

GUADALCANAL REPTOID   Reptilian humanoids reported from this Solomon Island. M. Boirayan feels they are extraterrestrials. They are manlike, but their skins are greenish-brown and scaly. They are bipedal and wear some clothes. They have been reported as flying through the air, perhaps aided by a device on their backs.
SIM: 40-41

GUINNESS LAKE MONSTER   A monster seen in this lake in the Wicklow Mountains in Ireland by two Estonian tourists in 2010. The lake, by the way, is made up of water, not Guinness.
information supplied by source at Radio Telefís Éireann

GURUMPA   A creature which looks like an ape and eats children in Nepali lore. There may be only one of them, perhaps invented to keep unruly children in order.

H

H'IK'AL   Literally meaning 'black man', this is a kind of demon animal in the legends of the Zozil Indians of Mexico.
CR: 2:1

HACKNEY RIVER FISH   In 1739 this alleged fish was caught by a surprised piscator in England. Its head resembled that of a pike, it had arms, paws, claws, four eyes and a crown on its nose. It was 6' in length.
FS: 2: 43

HADJOQDA   In Seneca legend, this is a human skin with nothing inside it. He hangs from a tree, protecting a nearby strawberry patch.
MAP: 18

HALENGAMER   An avian monster which was said to have lived in Lake Halen, Sweden, in days gone by.
www.cfz.org.uk: 11th November, 2010

HALF-WOLF   A creature described as half man and half wolf was reported in Brazil in 2008. It was stealing sheep and housebreaking.
www.iraap.org: rosales

HAMEL ANIMAL   An animal that no one could identify was on the loose in Hamel (Illinois) in 1933.
V: 232

HANNUSH   Wildmen in the lore of the Yagan Indians of Tierra del Fuego. They are perhaps identical with "black, wild men" on the island of which Charles Darwin wrote.
PM

HARFORD COUNTY BEAST   Lisa Mathis of Joppa (Maryland) saw from her window an anomalous creature with canine features about the year 2003. She couldn't identify it, but said it looked like something out of *Lord of the Rings*. She saw a similar creature on a television news bulletin about a year later.
www.wbaltv.com: news/3558698

HARRISON COUNTY MONSTER
A mysterious kind of monster has been prowling around the home of a man named Humphries in Harrison County, Kentucky. It attacked his dog, killing it the second time.

Despite this, local authorities seem to dismiss it as a domestic cat. However, it has is said to have a body like a lion, but the head is different. A possible photograph of the animal has been taken.
www.cryptozoology.com

HATUIBWARI   Reptilian beast or dragon believed in by the people of San Cristobal, Solomon Islands. It has a human head with four eyes, hands with claws, a serpent's body and bat's wings.
Wikipedia

HAWAIIAN MAGICAL DOGS   These appear in the mythology of Hawaii. Notable was Ku, who was able to change himself into a man and back into a dog again. Another was Ku-ilio-loa, a dog which could change from big to small whenever he wished. Another dog was Poki the wonder-dog, who was believed in in Oahu.
HLOH

HAWK MOUNTAIN DRAGON   This flying animal in Pennsylvania lore supposedly attacked a group of men in the Blue Mountains in the 18th Century.
MAP: 75

HEADLESS CREATURE   This creature was spotted at night by two anglers in the 19th

Century. It was about eight feet long and seemed somewhat doglike. Of course, it is probable the head was not discernable by the witnesses, whom it ignored.
SKM: 88

HEO BAC A form of pig reported by natives in Vietnam. It is whitish in colour with a short snout and white cheeks.
scienceblogs.com/tetrapod zoology: December 1st, 2010

HEO DEN Another kind of pig reported from Vietnam. It has a long snout and blackish colouration.
scienceblogs.com/tetrapodzoology: December 1st, 2010

HESSAFJORDEN SEA-SERPENT Brown serpentine creature with a dorsal fin which was seen off Norway in 1999 and 2001 and apparently at other times.
WW: 68

HO A bird in Chinese lore. It is supposed to resemble a pheasant. In colour, it is vivid green.
CMS: 227

HOG BEAR Smallish bear of American folklore, sometimes described as brown, sometimes as black. It has a snub nose and short legs.
V: 120

HOGA A monster said to inhabit Lake Themistitan in Mexico. Its head is pig-like and it has notable tusks. It can change colour, being variously red, green or yellow.
www.mythicalcreaturesguide.com

HOKEWINGLA In the mythology of the Dakota Sioux, a turtle to be found on the moon.
www.probertencyclopaedia.com

HOKKAIDO WOLF A Japanese subspecies of wolf (*Canis lupus hattu*) which received wisdom declares to be extinct, but of which modern sightings have been reported. It is one of two subspecies of wolf from Japan that are generally considered to be extinct, but that some cryptozoologists have hypothesized surviving. (see **HONSHU WOLF**)

The photograph on the right is courtesy Katuuya and Japanese Wikipedia.
Wikipedia

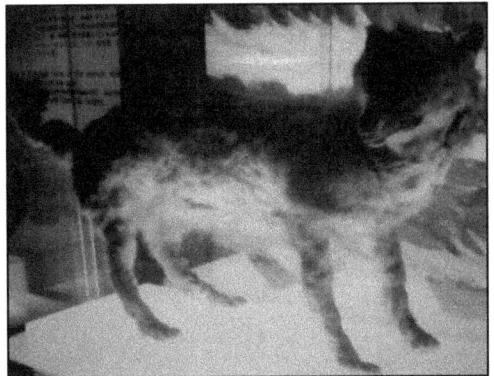

HOLLIS CREATURE A humanoid, perhaps of the Bigfoot kind, seen in Hollis (Mass.) in

1977. It rocked a camper van in which three people were sleeping. One emerged. He said later that the creature was brownish blond. The case was investigated by no less a luminary than Carleton Coon. Two women, each separately, also reported seeing the creature later.
www.wirenh.com

HOLLY CREATURE A creature seen in Holly (Michigan) in July, 2008. It was observed at night, investigating a dustbin. It had red eyes, hands rather like paws, long legs, a heavy lower body and a rat-like tail. It was grey in colour. It departed on all fours. The witness may have sighted it once more later in the month.
www.iraap.org: rosales

HOMBRE GATO A creature supposed to combine features of human and cat. It is the subject of belief in various parts of South America, notably Argentina.
Wikipedia

HONDDU WATER HORSE According to Welsh legend, this dwelt in the Honddu river. It would persuade people to mount it and then give them a rough ride.
paranormaldatabase.com

HONG KONG MYSTERY CAT Various unidentified cats have been reported from Hong Kong, like a leopard or tiger, but dark. An unidentified cat was killed in Hong Kong in 1989.
www.messybeast.com

HONSHU WOLF (see **HOKKAIDO WOLF**) A Japanese subspecies of wolf (*Canis lupus hodophilaps*) which has supposedly been extinct since 1905. However, modern sightings have been reported.
Wikipedia

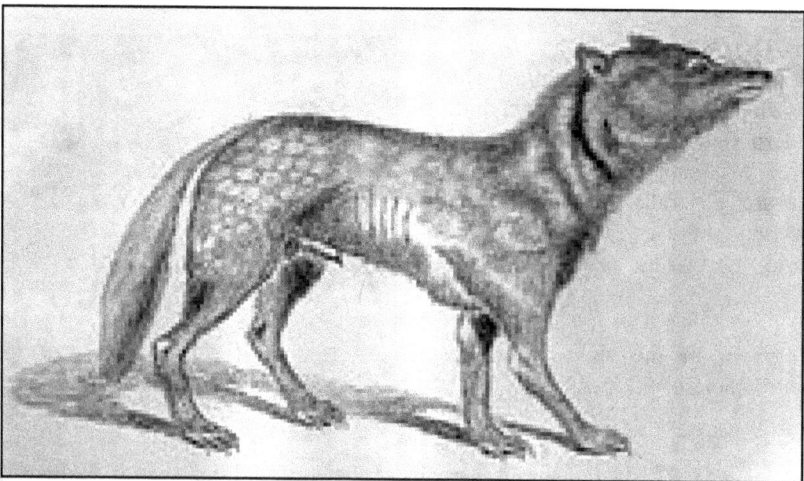

HOPPING HORROR A long-armed humanoid creature, apparently not hairy like a Bigfoot, has been reported in Vermont in modern times. There is some evidence that more than one of these creatures exists. As its name indicates, it proceeds with a hopping gait.
VMG: 18-19

HORIZON CITY CHUPACABRAS A creature that killed hens and rabbits, exsanguinating the hens, on a property in El Paso (Texas).
P 46: 13

HORSE-FACED SNAKE According to a Paiute Indian of Nevada, her grandmother once encountered a snake whose head resembled a horse's. It had a mane all along its back and whiskers. She seems to have seen one other specimen of this creature.
www.strangeark.com

HORSE-MAN OF BEDE In Mike Hallowell's blogspot, we learn that two people in a car in Bede, South Tyneside, were approached by a man in a hood and coat who banged (though not angrily) on the front of their vehicle. Then they noticed he had horse's legs and drove off in fright. Both driver and passenger were traumatised by the event. Two other instances of men noticed having hooves occurred in the same area.
www.cfz.org.uk: January 10th, 2011

HROSSHVALUR A mammal said to be found off the Icelandic coast, grey in colour, with a red mane.
WW: 37

HUACA MAMUL This beast is said to be found in a hidden lake between Aucupan (Argentina) and Chile. Its name means 'cow-stick', for it appears to float like a log, but do not be deceived. Approach it and it will rise up and attack you. It has a bovine appearance and its bellow is also like that of a cow.
cryptoflorid.ning.com: January 19th, 2000

HUAN Creature of Chinese lore. It is said to be able to survive on air alone, requiring no food.
CMS: 270

HUAY CHIVO An animal, part man and part beast, supposed by the Mayas to live in Yucatan, Mexico. There has been in recent times some confusion with the chupacabras.
Wikipedia

HU-CHIAO Creature of Chinese lore. No description is provided in the *Shan Hai Jin*, where it is mentioned.
CMS: 252

HUDSON RIVER MONSTER A monster of some sort has been reported in this river in New York state since 1610. The monster, nicknamed *Kipsy*, is now to be found on a mural in Poughkeepsie.
quadupdates.blogspot.com

HUIA A bird (*Heteralocha acutirostris*) formerly found on the North Island of New Zealand and generally regarded as extinct since 1907.

However, cryptozoologist Lars Thomas is fairly sure he observed one in 1991.
cryptodane.blogspot.com

HUMAN-WILDMAN HYBRID A woman who was kidnapped by a Wildman in China claimed he had made her pregnant and she had produced a son. A video recording of the supposed hybrid, now grown up, showed a male about 6'5" with a small head and caudal appendage. He could not speak.
World Journal (Taiwan): October 11th, 1997

HUMAN-WOLF HYBRID In Turkic myth, a bitch wolf gave birth to ten such

hybrids. One became the ancestor of the Ashina clan.

HUNTSVILLE CREATURE   This creature was seen at Huntsville (Utah) on 13th June, 2005. It stood like a human, but may have been hunchbacked. It was about three and a half feet tall. Its face was reptilian, but the body was hairy. There was a smell of bleach in the area.
www.aliendave.com

HUS KLAZOMENAIOS   A winged sow which was said in Greek legend to have devastated the island of Klazemenai west of Smyrna (now *Izmir* in Turkey).
www.theoi.com

I

IDAHO ANIMAL   An unknown animal killed in this state in 1909.   It had canine characteristics, but its species could not be identified.   It was much bigger than an ordinary grey wolf, its feet looked like a Newfoundland's and its sides and legs like a coyote's.   There was some curly black hair on its body.   The tail was both short and short-haired, not a canid's at all.
V: 215

IJAOE   Alternative name for the orang pendek.

ILLINOIS CATMAN   A car driving through Shawnee National Forest broke down in 1970. When the driver emerged, he was set upon by a humanoid with feline features.   Lights from another vehicle frightened off the assailant.
I: 264

INDAVA   A glowing flying creature reported on Umboi Island (Papua-New Guinea).   One was reported near Tawas Village.   P. Nation found what he felt to be a colony in 2006.   J. Kepas also claimed to have seen one.   There is speculation that the indava is a surviving form of pterosaur.
christian wiki

INDIAN GIANT SNAKE   A large serpent found in the Indus according to the ancient Greek writer Ctesias.   A flammable oil could be extracted from it.
www.cfz.org.uk: November 25th, 2010

INDIANA GIANT SNAKE   This snake, 40' in length, was reported from Muncie (Indiana) in 1895.
HR: 232

INDIANA MYSTERY ANIMAL   A strange creature, looking somewhat like a fox, but with little hair and a head too big and tail too long to be of the vulpine kind.   A number of

photographs have been taken.
cryptozoology.com: August 27th, 2009

INGANESS BAY MONSTER A huge creature that rose out of Inganess Bay at Mainland, the chief island of the Orkneys, and terrified two children during a violent thunderstorm around 1904. The creature was described as being too big for a seal and having a long, smooth neck. It had large, unblinking eyes and a head like a horse or cow.
NI: 118-120

INGLEWOOD POND CREATURE This animal was seen by two women at Alloa (Clackmannanshire), Scotland. It had the body of a horse and the legs of a dog. There were in fact two sightings, one in 1975 and one in 1997.
paranormaldatabase.com

INULPAMAHUIDA In Mapuche legend, a tree that can actually walk. It uses its branches to help it climb.
PM

IOWA FLYING CREATURE A huge flying creature was seen one night by a couple driving near Council Bluffs (Iowa). Its colour was olivish, but the couple disagreed on whether its face was that of a humanoid or a bat.
cryptomundo: August 27th, 2010

IRASBURG BIRDS Three huge unidentified birds with wingspans of about 15' seen in the 1980s at Irasburg (Vermont). There were at least five witnesses. One of them felt the birds were large enough to pick up a human. No feathers were noticed and the wings were totally stiff, like the wings of bats.
VMG: 105

IREDELL CAT Unknown species of light grey cat seen in Iredell County (North Carolina) in 2009. It was about 6' long.
V: 475

ISONADE This large sea-monster is said to be found in the waters to the west of Japan. Only its tail can be perceived. Because of this, one might ask if the tail in reality is most of the substance of the animal.
Wikipedia

ISPOLIN A kind of giant in Bulgarian mythology, it preceded men on earth. Ispolins lived on raw meat and fought dragons.
Wikipedia

IT A creature supposed to lurk around the Bake Oven Knob Shelter in Lehigh County (Pa). People seem unable to describe it. Do not confuse it with the Scottish creature of the same name.
CFZY 2009: 86

ITALIAN CREATURE In 1900 a gamekeeper in Piedmont turned to see a monstrous being with legs which resembled those of a hen.
JHS: 1:5

ITALIAN REPTILIAN This was reported by a number of witnesses in Padua in 1986. Reports are not terribly clear.
CFZY 2009: 149

J

JABOWAK This creature was seen in Frederick, Maryland, in 1870. It was tall and described as having a horrible face. The word is, of course, a corruption of Lewis Carroll's Jabberwock.
marksofmessengers.blogspot.com

JAPANESE KANGAROO A population of out of place kangaroos has been reported from Japan over the past seven years. Whether they are escapees, have been imported or are like the phantom kangaroos of the United States cannot be stated at this juncture.
Straits Times: March 8th, 2010

JAVAN TIGER The Javan tiger (*Panthera tigris soudena*) was declared extinct in 1994. However, droppings and paw prints indicate it might still exist. The specimen pictured was photographed in 1937.

This evidence has been found in the Meru Betiri National Park in Java.
Taiwan News: December 22nd, 2011 (website)

J'BA FOFI   In the beliefs of the Baka pygmies of the Democratic Republic of Congo, a gigantic spider the size of a human. Their webs can trap antelope and birds as they are strung between trees. Their dwellings look like the huts of pygmies. Reginald and Marguerite Lloyd may have seen one or something very like it in 1938.
FT 191: 54

JENTIL   A giant (plural *jentilak*) which occurs in Basque legend. They were hairy beings, tall enough to walk in the sea. They were said to have a certain culture, inventing metalwork. They were also thought responsible for megalithic structures. They finally became subterranean creatures with the advent of Christianity except for one, Olentzero, who locked his fellows in a cave and became a Christmas gift giver. Pictured right.
Spanish Wikipedia

JERICHO ROAD MAN-BEAR   This seems to be a creature in the well-known TNT area of West Virginia, from where the Mothman was reported. In 1966 one C. Lucas saw three man-bears in a field. They absquatulated on two legs.
WT: 63-4

JIQUE   In Cuban folklore, diminutive humanoids whose heads resemble frogs or are flattened. Their colour is black. They are hairy or scaly.
AE: 5

JOBRAN   Term used in the Himalayas for a Wildman with a reputation for eating humans.

JOHNSON SPACE CENTER MOTHMAN   At this NASA facility a worker, heading for his car after nightfall, saw a black figure on a building. It had what seemed to be a cape on its back and a pair of wings. The witness, who saw it in 1986, later discovered that there had been previous sightings.
www.mania.com: December 11th, 2010

JUMPING BEAR   A creature looking like a small bear sporting fangs and with a yellow chest. It could jump over considerable obstacles. It was said to have mutilated animals.
HSR: 2009

K

KAAIMAN This is the proper name of a specific mermaid in South African lore, said to dwell in the Buffelsjags River. She was seen by a number of persons in January, 2008, one of whom, Daniel Cupido, she nearly lured into the water, mesmerising him. She had red eyes and looked like a white woman with long black hair. Other witnesses were also present. www.iraap.com

KAJANOK A large creature about the size of a small boat observed in the waters off Greenland. It looked like a scorpion.
WW: 19

KANAWHA CREATURE A creature larger than a dog but smaller than a horse in the folklore of this West Virginia county.
WT: 19-20

KANGAROO VALLEY CAT This was seen by A. McDermott in 1964. It was greenish and had stripes. He described the head as like a bobtail. This valley, as its name would indicate, lies in Australia.
ABC: 339

KANSAS FLYING SERPENT A serpent of immense proportions was seen over Fort Scott (Kansas) in 1873. It seemed to form a circle around the sun.
HR: 230

KANSAS MONSTER SNAKE A snake 25'-30' long reported at Lury (Kansas) in 1933. It had glowing eyes. Its colour was described as "bluish-tan" which I cannot picture easily. www.cryptomundo.com: December 16th, 2010

KAP DWA There has been some doubt as to where the body of this giant ended up. Originally, Kap Dwa was a stuffed two-headed giant placed on exhibition in England. Two doctors and a radiologist could find no evidence that it was a fake. Conjoined twins, perhaps, they opined. F. Adey, who thought it a hoax, could not uncover any evidence of artificial joinings. One rumour had it that he had been brought to the United States. However, it is now known that he remained in Britain, ending up in the hands of a showman called Billy Hill.
FT 130: 54

KAPPIK An animal said in an early work to be part of the traditions of Greenland. It has been suggested that it refers to the wolverine. The wolverine, however, is not supposed to be found in Greenland.
BR: 15

KARAKONKOLOS   Beast man, not unlike Bigfoot, but apparently with the power of speech, in Turkish and East European folklore.

KAWTCHO   A kind of large hairy man which is believed in in Tierra del Fuego. It lives underground by day and preys on humans by night.
PM

KAYADI   A man-sized primate reported from Papua-New Guinea. It is believed in by the Siami and Amto tribes. It is about 5'5" in height and is a good tree-climber. It is bipedal.
AZ: 219; *FZ:* March 12th, 2011

KEMPENFELT BAY MONSTER   This bay at Barrie (Ontario) is supposed to house a monster.
Psican

KENEUN   The chief thunderbird in the mythology of the Iroquois.

KENTUCKY GOATMAN   Goatmen have been reported from Kentucky. One was supposed to have been at large in the 1840s in the vicinity of Tiline, Livingstone County. One was reported in Henderson County in the 1970s.
SKM: 89-90

KENTUCKY HYENA   This creature has been reported from Fayette County, Kentucky.
SKM: 119

KENTUCKY PRIMATE   A witness claimed to have been attacked in 1944 by a primate with a bushy tail which mauled him. It had grey fur and left five-toed prints. It was about the size of a human. Primates with faces that looked like a combination of pithecoid and human features, again with bushy tails, were reported in Kentucky in 1973.
SKM: 18

KENYAN MYSTERY ANIMAL   This appears to be a kind of giant elephant shrew, which was photographed in a camera trap. It is brown with a long nose.
www.americanmonsters.com: September 22nd, 2010

KHON PAA   In the 19th Century it was reported that this 5' hominid lived in Thailand. It had transparent skin, enabling the interested onlooker to see its internal organs. It had no knees: if it fell over, it had to crawl to a tree and haul itself up.
www.cfz.org.uk: November 19th, 2010

KING RAT   According to London legend, there is a rat in the sewers that is much larger than its fellows and also paler. He is surrounded by a rattish bodyguard and other rats keep silent when he hoves into sight. The pantomime *Dick Whittington* features a King Rat and one wonders if it is based on this legend. Although the pantomime is first recorded in 1814, a *Dick Whittington* puppet show was performed in 1668, shortly after the Great Plague, when

rats abounded and were much in public imagination. There is actually a species of rat called the king rat (*Uromys rex*) which is found in the Solomon Islands, but it bears no relation to its London namesake. The King Rat should not be confused with rat-kings, which is a term used to mean a group of rats whose tails have become inextricably intertwined. For the legend of Queen Rat, see the author's *Dictionary of Cryptozoology* (2004).
PL: 57

KING SNAKE A huge snake in the folklore of Chile. It is covered in black hair. The head is of great size. It is sometimes found guarding treasure in burrows.
Spanish Wikipedia.

KINIK In Alaskan legend, a sea-dwelling bear that is too large to drag itself out onto land.
V: 122

KIOWA FLYING SERPENT The Kiowa Indians are said to be familiar with flying snakes in Oklahoma.
HR: 251

KISHPIX A clawed animal, dangerous to man, in which the Yagan Indians of Tierra del Fuego believe.
PM

KISSING BUG In the year 1899 there was an outbreak of insect bites by an unidentified insect. It began in Washington (DC) and spread across the United States. The insect was never identified. It has, however, been given this curious name.
Masks of Mesingw: May 2nd, 2012.

KITCHENUHMAYKOOSIB CREATURE A strange creature washed up from a lake this part of Ontario. Its head was bald, its body long and hairy. Its face was like a cat's, its tail like a rat's. A shopkeeper claimed this kind of animal had been known for a long time, but the last recorded sighting had been forty years previously. It was discovered by a couple of nurses, who photographed it.
CSY: June 24th, 2010; *9News:* May 22nd, 2010

KO-GOK A kind of monster in the lore of the Abenaki Indians.

KOLOWOSI A friendly water serpent in the mythology of the Zuñi Indians.

KOREAN CRESTED SHELDUCK This bird (*Tadorna cristata*) may be extinct, but there is still some possibility that it is to be found. There have been reports from China, Korea and Siberia.

KRAGERØFJORDEN SEA-SERPENT This creature was described as horrible by those who saw it at this inlet in Norway in 1964. It was snake-like and bluish-green.
WW: 68

KUEI   Legendary Chinese fish. Do not eat its liver if you find one - it is poisonous as far as humans are concerned.
CMS: 199

K'UEI NIU   A kind of gigantic buffalo in Chinese lore. It may not have been an actual species of buffalo. Whatever it was, it existed as late as the 4th Century AD in Szechwan.
CMS: 230

KUPUA   In Hawaiian legend a kind of monster, not necessarily evil, who could assume any kind of body he wished.   Even after he was killed, he came back as a ghost, sometimes seen as a cloud in the shape of a dog.
HLOH: 56, 61

KUTABE   An ox-like or catlike creature in Japanese lore. Its face resembles a human's, but it has nine eyes and six horns.

L

LA NORIA HUMANOID   An unidentified skeleton discovered by Oscar Muñoz at La Noria, a town in the Atacama Desert, Chile. It was 15" in length, had nine ribs and a scaly body.
cryptozoologos.blogspot.com

LA PRYOR FLYING CREATURE   A grey flying creature with a wingspan of perhaps 8'-10', apparently featherless, seen at this Texas locale in February, 2009.
forteanwest.com

LAGO TODOS DE SANTOS MONSTER   This South American lake is said to harbour a monster with a tail that ends in a spearhead.
PM

LAKE BANYOLES MONSTER   This lake in Gerona, Spain, was said to harbour a monstrous denizen. Legend describes it as a dragon which has been there since at least the 8th Century. It had thick scaly skin with spikes protruding from its spine. Its breath befouled the air. It had large wings, but was too heavy to fly. Reports of a monster in this lake are recorded in the 19th Century and the early part of the 20th. It could emerge from the water and on one occasion it is supposed to have crossed the road. On May 26th, 1913, a boat was wrecked on the lake and local gossip blamed the monster.
criptozoologos.blogspot.com

LAKE BIWA CENTIPEDE   This appalling insect allegedly lived in Japan in the 10th Century. It was killed by a Japanese general.
www.cfz.org.uk: September 28th, 2010

LAKE BÖRRINGE MONSTER   A monster or monsters was/were seen vaguely in this lake

in Sweden in 1988.
WW: 88

LAKE BUENOS AIRES SERPENT *see* **Lake General Carrera Serpent**

LAKE BULL   Creatures called lake bulls, looking like bulls or cows, are reported from a number of lakes in Patagonia.
PM

LAKE CAMI MONSTER *see* **Lake Fagnano Monster**

LAKE CAVIAHUE CREATURES   Mystery animals resembling bulls and horses have been reported from this Argentine lake.
PM

LAKE CHELAN MONSTER   A photograph was taken of the head and neck of this monster sticking up in this lake in Washington state in 2007. The photographer claimed he had been able to make out some of the monster's body under the water.
www.cryptozoology.com: July 8th, 2007

LAKE CHANY MONSTER   There are reports of a monster in this Siberian lake. An angler fishing in the lake hooked something and was pulled in. His remains have not been found. Others have disappeared in the lake as well.
FT 267: 8

LAKE CISNES ANIMAL   This Argentine lake is reported to be home to a strange animal, like a cow above and a horse below or like a cow above and an unidentified animal below.
PM

LAKE CLINCH MONSTER   Indians had a tradition of a monster in this Florida lake. There was a sighting of a 30' long creature from Frostproof in 1907. A man called Mallet disappeared when swimming in the lake in 1926 and, when his mangled remains were discovered, it was suspected he had encountered the monster.
cryptozoo-oscity.blogspot.com: April 22nd, 2010

LAKE COLERIDGE MONSTER   A monster was reported in this New Zealand lake in the 1970s. It was given the nickname "Lakey". A wolf-like head emerging from the water was spotted by two women in 1975. Earlier a fisherman's bloodstained boat was found, without the fisherman in it and the idea was mooted that the monster had dragged him into the water. The creature may be able to go on land, for sheep have been reported missing from nearby. In 1979 the monster fixed its eyes on a boatload of fishermen.
www.cfz.org.uk: 13th November, 2011.

LAKE COUNTY ANIMAL   An animal shot by one Archie McMath in this area of California in 1868. It was built differently from a panther, by which I presume a puma is

meant. It was heavier in the front, of a yellow colouration, with black stripes, resembling a mane, on its shoulders and front.
V: 156

LAKE DUSIA MONSTER There is a vague story of a serpentine creature in this lake in Lithuania.
WW: 122

LAKE FAGNANO MONSTER This lake in Tierra del Fuego is also known as Lake Cami, as it is shared between Argentina and Chile. It is supposed to harbour a monster called *Fañanito*. It may be that the story derives from sightings of beaver, which are found in the lake.
PM

LAKE GENERAL CARRERA SERPENT This lake is shared by Chile and Argentina. In the latter country it is called Lake Buenos Aires. A serpentine creature, 15'-18' long, has been reported here.
PM

LAKE HOPCATONG MONSTER In the lore of the Delaware Indians, a monster had once dwelt in this lake. It had huge antlers or horns and the texture of its body resembled an elephant's. It hasn't been seen alive for about two hundred years, but some settlers are said to have seen its body underwater.
MNJ: 77-8

LAKE KAMPESKA MONSTER A creature of considerable proportions, compared with a sea-serpent, was seen in this lake in South Dakota in 1886.
HR: 233

LAKE KRANKE MONSTER What appeared to be a creature like an upturned boat was seen moving back and forth on this Swedish lake in 1979.
WW: 89

LAKE METELYS OTTER A giant otter, sighted just for a moment in 2002 in this lake in Lithuania. Only the head was seen.
WW: 121

LAKE MOUSTOS MONSTER A monster is supposed to lurk in this lake in Greece. While no one claims to have seen it, many have heard a booming sound ascribed to it.
HG: 19

LAKE PEPIN MONSTER This monster is supposed to haunt Lake Pepin (Minnesota). There have been reports since 1871. It proceeds by undulation and is of a greenish-grey hue, if accounts are to be believed.
AZ: 275

LAKE PERUCA MONSTER There have been a number of reports of a monster in this lake in Croatia. Its body is supposed to be fat and its head large.
AZ: 275

LAKE POWELL MONSTER Two anglers reported a creature in this lake in Arizona on July 31st, 2011. Two humps were discerned. It was also seen on 1st August. Both sightings occurred near Lone Rock. One estimate of the length was 50', but such estimates are frequently exaggerated.

These were not the only sightings. We refer you to an attack on Eric Padgett in 1958. Padgett was sure it was a mammal.
www.examiner.com

LAKE SOMIN MONSTER Reports of this creature in a lake in the Ukraine occur. Animals and humans are its victims and it makes strange noises at night. It is said to have a snake's head and a crocodile's body.
cryptoflorid.ning.com: August 23rd, 2009

LAKE TAR MONSTER Indians claimed this lake in Argentina boasted monstrous animals with long necks but no manes.
PM

LAKE VILLARICA MONSTER This Chilean lake is supposed to harbour a monster, but sightings are inconclusive.
PM

LAKE VOMB MONSTER An animal with a long neck and a small head was seen in this Swedish lake by a couple in 1991.
WW: 89

LAKE WASHINGTON OCTOPUS This lake is said to harbour octopuses (or octopodes, if you prefer the older plural) which have been reported up to the present day.
SM: 59

LAKE WILLIAMS MONSTER A teenage couple claim their car was attacked by a huge mystery reptile near this lake in Michigan in 1973.
www.americanmonsters.com: May 3rd, 2010

LAKE WILSON MERMAID Several Hawaiian school students reported, in this lake, a green lady with seaweed-like hair and long green claws. This occurred in the 1950s.
I: 31

LAKOOMA Sea-monster said to be found off Tierra del Fuego.
PM

LANCASHIRE CREATURE   Monster reported from an unnamed lake in Lancashire. The witness, an angler, took video footage of the creature.
P 56: 11

LANCASTER COUNTY CREATURE   A bipedal creature which two brothers reported seeing in Pennsylvania. It had curved horns, a white mane, ferocious fangs and was coloured grey. This or a similar creature attacked a farmer who defended himself with a scythe, which the creature seized, whereupon the farmer beat a hasty retreat. The creature (or another of the same kind) stole two geese from another farm.
SI: 143-4

LANGANA CREATURE   Off the coast of the Greek island of Zacynthos near Langana, this creature, half fish and half serpent, is believed to infest underwater caves. It can also travel on land.
HG: 70

LAPHIATE   A two-headed serpent in the folklore of the Greek island of Naxos. Two-headed snakes that are freaks of nature are sometimes found, perhaps giving rise to the legend.
HG: 60

LAREDO MONSTER   An unidentified creature washed up on the shores of Cantabria, Spain, in 1974. It was perhaps a rotting basking-shark.
Criptozoología en españa: March 15th, 2012

LAUGHING OWL   A New Zealand bird (*Sceloglaux albifacies*) listed as extinct, but there is the possibility that belief in its extinction may be premature.

It was also known as the *hakoke, whekau* and *white-faced owl.* There was a decline in this species after 1840. The last confirmed sighting of one was in 1914.

However, evidence of its continued existence consists in possible sightings, fragments of eggshell and some Americans hearing what sounded like its calls in 1985.
CFZY 2008: 145-8; *Wikipedia*

LAWRENCEVILLE BIRD   A gigantic bird spied over this area of Pittsburgh (Pa) in May, 2010. The creature gave a loud screech. The sighting occurred at night.
JHS: 1:2

LEAVENWORTH FLYING SERPENT   In 1875 a snake with spots and wings was

allegedly captured by two boys in this part of Kansas. A woman of Leavenworth had previously reported seeing a flying snake.
HR: 249

LEHARVEN Name used for the nattravnen in Blekinge.

LEVERETT ANIMAL An unknown striped animal startled horses pulling an equipage at Leverett (Mass) in 1899. A hunt was mounted to track it down, but was unsuccessful, although tracks which none could identify were found.
V: 319

LINGETA A sea-monster supposed to live off the coast of Epirus, Greece, near Mount Himaras. No description is available, due to the fear of sailors to furnish it.
HG: 70

LIOKORNO A large serpent in Greek folklore. It is said to have one horn.
HG: 59

LION-SIZED FELID A creature rumoured to be found in Papua-New Guinea.
AZ: 219

LIVING WOOLPACK This was reported in Ireland to an informant of Lady Gregory, folklorist and mythologist.
MAI: 100

LLUHAY In the folklore of Chiloe (Chile), an immortal reptile, silvery in colour. It has tusks and a liking for potatoes.
Spanish Wikipedia.

LLUL-LLUL A sea-dwelling creature like a cat with a long tail in the beliefs of the Mapuche Indians of Chile.
PM

LOBO-TORO Mapuche Indian legendary animal having characteristics of both wolf and bull.
PM

LOCH TINKER MONSTER This lake in Stirlingshire (Scotland) was home to a monster said to have devoured a tinker, from which act the loch obtained its name.
MS: 112

LO-LO Legendary bird of Chinese lore. This term is also used to mean a tiger.
CMS: 236

LONDON HAIRY CREATURE A short hairy creature reported from the Hollow Ponds

area of Epping Forest in November, 2008.
www.paranormaldatabase.com

LONG LAKE MONSTERS Two huge reptiles, which looked like lizards with alligators' heads and serrated dorsal ridges were seen fighting in a stream near this lake in North Dakota in 1883. Do not confuse them with the monster of Long Lake (Wisconsin).
HR: 233

LONGANA A beautiful woman having the legs of a goat, rather like a female faun, in Italian folklore. Belief in the longana is found in the region of Cadore.
Italian Wikipedia

LONGMA A horse covered with scales in Chinese mythology. Its name means *dragon-horse*. It is called *ryuma* in Japanese.
Wikipedia

LOOE MONSTERS Two monsters looking like dragons were reported off the Cornish coast in 1949.
www.paranormaldatabase.com

LOUGH CIME MONSTER The precise location of this lake in early Ireland cannot be determined, but it was supposed to harbour a monster which swallowed a man. Happily, a saint made it throw up. The man was ejected alive.
MAI: 73

LOUGH LEANE MONSTER There are vague suggestions that there is a monster in this lake in Kerry, Ireland.
MAI: 55

LOUGH LOOSCANAUGH MONSTER There is a tradition of a monster or monsters in this lake in Kerry, Ireland.
MAI: 56

LOUGH NEAGH MONSTER This is Ireland's largest lake. A long black creature was spotted in the water here in 2006.
www.paranormaldatabase.com

LULU A creature compared with a giant or ogre by African pygmies. They believe it can change itself into a huge serpent.
FZ: April 19th, 2011.

M

M BOI TU'I In the myths of the Guaranis of South America, this is a huge serpent with a parrot's head. It protects water dwelling creatures.
Wikipedia

McHENRY BIRDS Giant birds that flew in a way reminiscent of bats were observed over McHenry (Illinois) in May, 2010.
JHS: 1:2

MAD DOG An unidentified creature that might have been a dog, wolf or something else. Whatever it was, it had very fierce teeth. It was to be found in Kanawha County (West Virginia). It would attack riders on horseback, assaulting the horse first.
WT: 20-21

MAGDALENA ISLAND APE-MAN Magdalena Island is one of the the Guaitecas Islands in Chile. Here, an ape-man was allegedly seen in the 1940s.
PM

MALVERN CAT A felid was seen in a cornfield near Malvern in 2000. The witness was sure it was either a leopard or jaguar.
GW: 36

MAMMOTH With regard to the continued survival of this creature, G. McIsaac, using Canadian Indian sources, says it is reputed to be still in existence. It is said to show great hostility to man. It inhabits caves in winter. There was an actual sighting of one in 1965.
BH: 55-57

MAN BEAST OF DARIEN A large humanoid which walked upright, standing 6' tall, seen in Panama. The story of the sighting was told in 1920, but the sighting itself cannot be dated.
www.bigfootencounters.com

MANAC CREATURE A creature reported from Manac County, Illinois, which seems to be similar to the Tuttle Bottoms Monster.
www.cryptozoology.com

MAN-BAT Two musicians, father and son, were driving home from La Crosse (Wisconsin) when a furry humanoid, 6' in height, with fangs and what they estimated as a 10' wingspan, landed on the bonnet (or, as it is known in transatlantic parts, the hood) of their car. It flew upward with a scream. They experienced a subsequent illness which lasted some days. They lived in a house in the country and for some time they experienced pounding on the door and rattling of the doorknob, as though something were seeking ingress.
MW: 37-38

MAN-BIRD   This 4' tall creature, which seemed to be part man and part bird, was seen in Rio Grande (Texas), according to local gossip, in November, 1975.
MT: 94

MAN-DOG   A creature reported from Chile, said to have had the head, face and arms of a dog and a human body, including the legs. It was reported in the vicinity of La Serena and Coquito.
criptozoologos.blogspot.com: September 7th, 2011

MANITOBA RIVER MONSTER   In a report of 1885, one J. Bryan was fishing in an unnamed river, when he saw a large unidentified animal, with a head like that of an hippopotamus.
cryptosup.com: September 18th, 2010

MAN-SIZED OWL   A witness at the age of twelve, about 1997, saw a huge owl near Norfolk (Va). It was at least 5' tall.
cryptozoology.com: June 9th, 2009

MAN-SIZED RAPTOR   Large bird, 8' tall, seen near Rangerville (Texas) in the 1990s. It was sitting on top of a telephone pole
MT: 11

MARABECCA   In Sicilian folklore, a creature that lives in wells and reservoirs.   No description is to hand.
Italian Wikipedia

MARIPILL   Among the Mapuche Indians of South America, these creatures are described as large and unsightly. Their backs are serrated and they have clawed hands.
PM

MARMAELER   The children of mermaids and mermen in Norwegian lore. It was believed that sometimes they had been captured by fishermen.
NM: 246

MARTELLO MONSTER   An unidentified submarine creature that rent fishermen's nets at Seaford (Sussex) in 1968.
www.cfz.org.uk: December 27th, 2010

MARTIAN CREATURE   The English newspaper the *Sun* (March 3rd, 2010) produced a NASA photograph of a section of Mars showing what appears to be an animal resembling a gorilla. While N. Cooper, who studies NASA photographs, is sure it is alive, it might be a geological feature.
FT 270: 25

MARYLAND CREATURE   A teenager claimed to have seen this creature in her bedroom in

2005. It looked like a small mummy, its skin was black and it was very boney. It was hunched, bald and wrinkled. The witness's mother had noticed a strange smell in the room for some days before the sighting.
About.com: *paranormal phenomena*

MATA A monster in Scottish lore which had four or seven heads, 1400 feet and is described as a pig-monster.
GNB: 188, 190

MATAU GIANT This species of New Zealand hominid is thought to be found in the region of Lake Wakatipu (Otago).
nzcryptozoology.ucoz.com

MATTAWA CREATURE A bipedal creature with a large, white, round head and an apparently thin layer of fur, seen at this location in Washington state in 2009.
ufoinfo/hominid: 2009

MAZINAW LAKE CREATURE A creature in this Ontario lake said to look like a giant eel. However, it may be no more than a very large fish.
cryptoflorid.ning.com: October 7th, 2009

MECHANICSBURG CREATURE A humanoid of diminutive stature which had feet seemingly resembling a bird's seen by a single witness at night in Pennsylvania in 2009.
ufoinfo.com/humanoid: 2009

MEDDYBEMPS HOWLER A term once applied to certain black-haired hominids in Maine. Indian women described them as being 'two fathers high'. Their howling was in fact more like singing.
www.bigfootencounters.com

MEDEK The legends of the Gitksan and Wat'suwet'en Indians of British Columbia have a story of a city called Dimlahamid (or Dzilke) which underwent a number of catastrophes. One was caused by a lake monster called the Medek. The Indians seem rather unsure of what the Medek was. It was a kind of wolf, puma or bear in various versions, which would indicate they may have had no real memory of its actual appearance.
DFD

MEDINA RIVER FLYING CREATURE Seen at this Texas locale, it was a large flying creature with a wingspan of 15' or more which the witness asserted bore no resemblance to a bird. It was reported on August 11th, 2009.
phantomsandmonsters.wetpaint.com

MEDWAY CREATURE An eel-like creature, said to have been impervious to gunfire, reported from the Medway river in Kent in the 1990s.
P 55: 73

MEDWAY MONSTER One Richard Mann reported a black, circular object, perhaps living, near Rochester Bridge on the River Medway (Kent) in the 1980s. In 2007 a 30' long beast was reported in the same area.
P 55: 73

MELUN PRIMATE According to a report in the *China Mail* (October 25th, 1947) this area south of Paris was visited by a gorilla-like creature wearing a red overcoat and shoes. It was swinging from the trees. This snippet was dug up by the indefatigable Richard Muirhead.
www.cfz.org.uk: December 5th, 2010

MER-COCK This strange creature was supposed to have appeared near Weymouth (Dorset) in 1457. It looked like a rooster, was coloured like a pheasant and crowed three times. Then it "vanished awaie".
www.darkdorset.co.uk

MERMAID A modern alleged mermaid sighting took place at Kiryat Yam, Israel. This mermaid would only appear at sunset. She was at first mistaken for a sunbather by witness Shlomo Cohen. Then she dived into the sea and a number of people saw the tail. A reward of $1 million has been offered for a photograph of the creature.

Greek folklore knows of a large mermaid called the Gorgona, sometimes beautiful, sometimes hideous. She was Alexander the Great's sister. The tradition was she would ask sailors if her brother still lived and, if they answered negatively, she would sink the ship. Alexander's sister was also known as Thessalonike. In Thrace the mermaid was Alexander's mother and called Phokia.

The Mermaid was the name of a tavern in London which existed from 1464. This shows how clearly the creature was embedded in English consciousness at the time. Some English

mermaids are said to dwell in inland bodies of water. There is supposed to be a mermaid at Mermaid's Pool, near Hayfield (Derbyshire). If you see it, it is said you will become immortal. At the Black Mere on Morridge Moor in the same county, a dangerous mermaid who will pull you into the water where it is supposed to live.

A childlike merman was reported in Lake Superior in 1782. According to Indian legend, mermaids with fish tails were to be found in the Merced river of California. Benjamin Franklin (1705-1775) claimed that a mermaid was seen off Bermuda. It was the size of a 12-year old boy and had a fish's tail.
HG; www.michigansotherside.com; *Wilson:* Cap. V; *Pennsylvania Gazette:* April 29th, 1736

MERMAID-HUMAN HYBRID Danish folklore furnishes us with a character who was the son of a mermaid and a human. His name was Bondevette.

In Ireland, a number of families are said to be of mixed human-mermaid descent.
NM: 324; *MAI:* 127

MEXAXKUK A horned serpent in the lore of the Delawares and other Indian tribes. It could sometimes turn into a man. If, in this guise, it made a girl pregnant, she would give birth to snakes.
MAP: 13

MIALUKA The Kansa Indians of Kansas and Missouri believe in giants with big heads and long hair and call them by this name.
AE: 168

MICHIGAN CATMAN A feline humanoid which attacked a car in Niles (Mich) in 1969. It thumped the car with clawed fists, breaking windows.
I: 264

MICHIGAN CREATURE Mysterious animal seen by hunters near the Canadian border, showing characteristics of polar bear, wolf, ape and big cat. It had black skin on its face and stomach, but otherwise appears to be white.
FZ: October 7th, 2012

MICRO-MERMAID In Makati City, Philippines, a young girl named Sophie captured fish in a bag and discovered one to be an extraordinary creature. It was almost transparent, so she thought at first she must be imagining it. It had large eyes, small nostrils and sharp little teeth. It had six arms which had hands which seemed to have three fingers each. Unfortunately, other fish in the bag quickly ate the unfortunate creature.
www.americanmonsters.com: December 20th, 2010

MILWAUKEE BEASTIE An unidentified animal, apparently showing feline and/or canine characteristics, seen near the Milwaukee River in 2008. Three photographs were taken.
cryptozoology.com: March 12th, 2008

MINNESOTA MOTHMAN Large creature observed outside Stewartsville (Minnesota) on June 27th, 2007.
www.cryptomundo.com: August 11th, 2007

MIRKA Term for a sort of Wildman used in the Himalayas.

MISSOURI FLYING SERPENT Tales of a flying serpent that had flown over a Missouri steamboat were current about 1857. It was said to be undulating, which is not the way snakes move on the ground. It was also said to be breathing fire.
HR: 231

MR DAVY'S MONSTER This curious creature was first seen in London in 1878. It was 2' high and 2' long. It seemed to lack a body between its forelegs and its hind legs. It was hairy and the head was like a boar's.

Mr Davy, the owner, seemed to be in full control of it. No one knows what the beast was, but it seemed to inspire horror in all who saw it, even though it appears to have been quite harmless. There was a rumour that it had been purchased in France, but that Mr Davy himself had no notion of what it was.
PL: 54

MIZUCHI In Japanese lore, a sort of dragon that has watery associations.

MOEHAU A hominid of human size in the traditions of the Maori. They are said to be shaggy creatures with talons. Watch out if you are in the vicinity of the Haast River: that is supposed to be one of their favourite places. There is also a Moehu Mountain. The state of a headless prospector who had been partially eaten in 1882 was ascribed to their endeavours.
nzcryptozoology.ucoz.com; *CFZY 2008:* 164

MOLDA SEA-SERPENT This sea-serpent was seen off Norway by Commander de Ferrys in 1746. Its head was like a horse's, with a long white mane. Its colour was greyish, but its snout was black.
MN: 249

MOÑÁI This is a huge snake, of which there is but a single specimen. It is believed in by the Guarani Indians of South America. It has a set of antennae and hypnotic powers.

MONCLOVA QUADRUPED Manlike quadruped with grey, smooth skin seen in Mexico in April, 2010. In fact, several of the creatures were reported. One was seen to dive into a pond.
JHS: 1:1

MONKEY-DOG A strange creature which showed characteristics of both monkey and dog was seen by a woman at Youngwood (Pa) twice in 1973, once in her front garden, once on her doorstep. It had a hump, beautiful brownish-black hair and a wide, ringed tail a foot long.
SI: 198-9

MONKEY-HEADED BIRD   This strange creature, 3' long and 8' broad, was supposedly captured in Hong Kong.
Perth Mirror: July 17th, 1937

MONSTER OF KOBYAKOVO GORODISZHE   The Kobyakovskiye Caves are situated in the North Caucasus.   Local lore claims the caves harbour a monster of unspecified description.   Persons and animals in the area have disappeared.   In 1949 a party was sent in as part of a Soviet plan to build a military centre there.   The first party returned unscathed, but the second party vanished.   Investigations found the bodies of two of its members, one headless, the other excarnated.   The military installation was built elsewhere.
JHS: 1:4

MONSTER OF MOUNT PILCHUCK   A huge winged creature, about 8' tall, with leathery wings was alleged to have attacked two carloads of people at this locale in Washington state in 1981.   Police and army were called in.
www.cfz.org.uk

MONSTER OF WINTERFOLD   In 1967 a couple driving in Surrey were troubled by their car's breaking down.   When the driver emerged to examine the engine, they were struck by a horrible odour, not unlike that of the stink bomb favoured by prankish schoolchildren.   On getting back into the car, he saw a face looking through the window.   It was white and he could not see eyes, nose or mouth.   Though the head was white, the body was dark.   The driver estimated the height of the creature at 4'6".   He said the body looked bell-shaped.
MAL: 302, 304-7

MONTAUK MONSTER   A strange animal whose carcass was found on the beach at Montauk (NY) in July, 2008.   No one could identify it.   It might have been an experiment from the nearby Plum Island Animal Disease Center.   However, many, including cryptozoologist Darren Naish, regard it as a raccoon.   W. Wise (Stony Brook University) said the creature was probably a fake.   However, on January 11th, 2011, a similar beast was found on Long Island which was said to bear no resemblance to a raccoon.   Yet another similar beast was discovered by L. Ingmann on Silver Sands (Connecticut).
www.montauk-monster.com; *Wikipedia*

MOORE LAKE MONSTER   This lake was caused by damming the Connecticut River in Vermont in 1957.   A monster from which a red glow issued was seen there in 1968.   Two glowing red spots approached a boat containing anglers.   They saw a kind of mound behind the eyes, with perhaps a larger part of the creature behind that.   The anglers survived, but later many dead fish, presumably the monster's victims, were found.
VMG: 90-91

MORGAN'S RIDGE MONSTER   Morgan's Ridge is to be found in the town of Rivesville (West Virginia).   The Monster seemed to have been seen chiefly around the 1920s or 1930s or perhaps earlier.   A family named O'Dell are said to have suffered from its depredations on their sheep and to have eventually shot it (they heard it scream), but they could only make out its shadowy form in the dark and they did not find its body after the shooting. Frank Kokul, a Croatian immigrant, came face to face with the Monster in 1929 and fought with it. It seemed phantasmal for his blows did not seem to affect it and, despite its attack, he received no injury. The beast vanished.

Sightings of this monster have spanned a number of generations.
WT: 1-5

MORONA COCHA SERPENT   A married couple lived in a house on stilts in the Morona Cocha lake in Peru, when in 2009 a monstrous serpent in the lake dragged an island across the water into their house. Happily, they escaped.
P 46: 60

MORTON GROVE ANIMAL   Animal seen at night in a suburb of Chicago. It crept across the road. It was 2-2.5' tall at the shoulder and a superabundance of fluff disguised its true shape. Its bushy tail was at least twice as long as its body. The animal was reported in 2005, but the incident occurred some years earlier.
cryptozoology.com

MOSCOW GHOST CAT   This cat is said to appear at midnight in Tverskaya Street in the Russian capital.
indrus.in/articles: November 8th, 2010

MOSES LAKE MONSTER   This lake in Washington state was the site of a sighting of a monster that appeared reptilian in 1992.
SM: 63

MOUNT HOOD CREATURE   A creature of greyish colouration, 10'-12' tall and very thin seen by two witnesses at this Oregon location.
www.iraap.org: Rosales

MOUNT VERNON MONSTER   The howls of this monster were heard in the region of Mount Vernon (Va). A forested area near Southwood seems to have been the source. This occurred in the 1970s-1990s. Thelma Crisp said she had seen the creature making the sound. It was heavy, bipedal and about 6' tall. It was said to eat food left out for it. A police search with a helicopter failed to reveal it.
MAV: 73-75

MOUNTAIN DEVIL   A creature reportedly found in New Mexico. However, it seems likely the term is applied to a number of different animals. One description compared it to a jackass. Another said it was a plantigrade with ursine hair, but its legs were long and straight, its back straight also, while the head was like an hyena's. In Australia, the reptile the thorny devil (*Moloch horridus*) is sometimes referred to as a mountain devil, but its existence is not in doubt.
V: 389-90

MUCK MONSTER This peculiar creature has been spotted in a video recording of Lake Worth (Florida). It has defied identification.
cryptoflorid.ning.com: August 21st, 2009

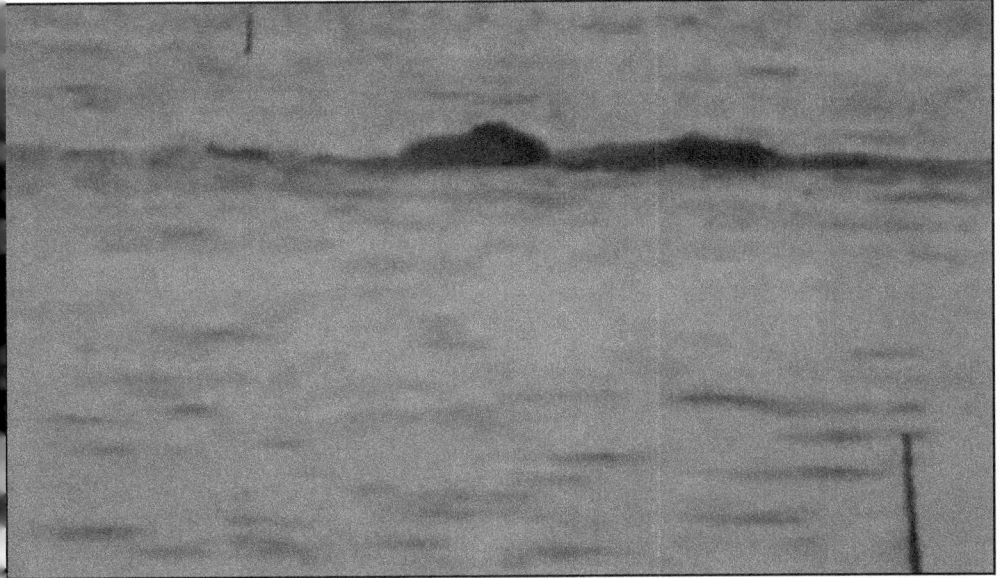

MUGWUMP This rather vague term seems to be applied to a number of monsters, but actually it is merely an Algonquian word meaning 'eminent man'.

MUSHVELI A sea-creature of Icelandic folklore. The name means 'mouse-whale'. There was a possible sighting in 1964, in which the witness claimed it had ears resembling an elephant's.
WW: 36-7; AZ: 291

MUWA A somewhat simian cryptid, said to have wings like a bat and hooves or lizard feet, allegedly found in the Philippines.
Wikipedia

MWONO A kind of Wildman in which the Alakaluf Indians of Tierra del Fuego believe. They claim to have seen his tracks. His name means a snow man. Although he is thought to be an individual, this may be because only single members of the species have been observed. He lives on the mountains and glaciers, not venturing into inhabited areas. However, you enter his domain at your peril.
PM

MYSTERY FISH A fish whose like its captors had never seen was allegedly captured in 1786 and brought to Gravesend (Kent).
www.kentmonsters.blogspot.com

MYSTERY INSECTS   An influx of these, resembling (but presumably not identical with) cockroaches landed in Woolwich in the 1990s. They seemed to show a preference for the Globe Industrial Estate. Their origin is unknown.
K: 329

N

NABAU   A snake with a dragon's head and seven nostrils in the folklore of Borneo. It is supposed to be able to attain a length of 100'. On the 21st January, 2009, a photograph was taken from a helicopter over the Baleh River in Borneo, apparently showing a gigantic snake, which locals identify with this animal. Another photograph has also been produced.
dailymail.co.uk: August 18th, 2010; *news bizarre.com*

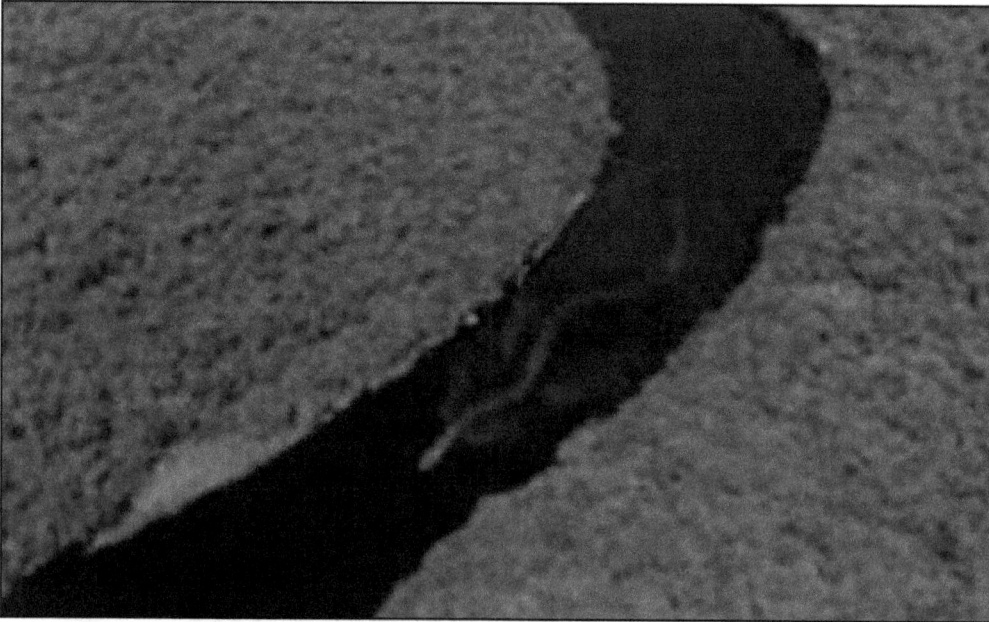

NAMIBIAN ANIMAL   Unidentified animals, which kill other animals, exsanguinating them, have been reported from Namibia. One witness claimed he saw an animal that looked like a dog, but was too big. It had a black or brownish head, but the rest of its body was white.
www.cryptosup.com: October 14th, 2010

NANORLUK   In Alaskan legend, a large bear. It will chase humans and swallow them whole.
V: 122

NANT GWYNANT WILDMAN   At this area in North Wales at an unspecified long time ago, a red-haired hominid was raiding livestock and was seen to flee on all fours. Once it tried to enter a house through a window and a woman inside cut off its hand with a hatchet. The creature made off and did not return.

NAPADOVKA CREATURE A white animal, resembling a kangaroo, seen in this part of the Ukraine in 2009. Animals in the region had been previously exsanguinated.
ufoinfo/humanoid: 2009

NARGER A living stone in the legends of the Australian Aborigines of Victoria. It supposedly dwelt in a cave behind a waterfall in Mitchell River National Park.
www.cfz.org.uk: October 6th, 2010

NARROW LAKE MONSTER A monster with a snakelike head was reported in this Michigan lake in 1886.
HR: 233

NATTRAVNEN A dark flying monster which looks like a bird and will devour the lone wayfarer in the folklore of Sweden. In Blekinge it is called the *Leharven.*
www.cfz.org.uk: November 11th, 2010

NAWAO Large wild people of Hawaiian legend. They do not seem to have been considered human. They avoided men and had an inclination to eat bananas. They have now disappeared.
HM: 32

NEBRASKA CRYPTID This was seen on 13th June, 2006, by Mary Ann Carta, who managed to take a photograph of it, partly hidden, which renders it difficult to identify.
cryptomundo.com: June 14th, 2006

NENNORLUK A sea-creature in the folklore of the natives of Labrador. It has a white back, huge ears, walks on the sea floor rather than swims and is able to submerge. It favours seals for its diet, but will also eat humans. There was a sighting at Cape Mogford in 1847.
The St Johns Telegram website: October 10th, 2008

NESOPHONTES A rodent, supposedly extinct from the 17th Century. However, there is some evidence the Haitian and Cuban species still exist.
cryptomundo.com

NEVADA FLYING SNAKE This was reported has having been seen in Virginia City in 1885.
HR: 250

NEW BRITAIN DINOSAUR The island of New Britain, part of Papua-New Guinea, has adjacent islands. On two of them, Ambungi Island and Alage Island, a dinosaur-like creature has been reported from the 1990s. It has been compared with a Therizinosaurus.
Paranormal About.com

NEW CUMBERLAND CREATURE An animal resembling a white bear was seen twice by the same witness at New Cumberland (West Virginia). On the second occasion, he and a friend were actually chased out of some woods by the creature, which then screamed.
WT: 71-2

NEW MEXICO DINOSAUR This animal was reported in New Mexico by a bus driver. He said it resembled a picture of a diplodocus he subsequently saw in a book.
AZ: 294

NEW ORLEANS LIZARD MAN Around February 2010 a dark-hued creature, its skin covered with what resembled algae and possible gills on its head, a jutting forehead and yellow eyes, was reported in the New Orleans area by a trucker.
graelianreport.com: February 11th, 2010

NEW SOUTH WALES CREATURE Huge creatures of the southerly highlands of New South Wales were allegedly sighted in the 19th Century. In Aboriginal tradition a certain billabong in the Lismore region was home to these animals.

It has been suggested that these creatures were surviving diprotodons and, if so, their current existence cannot be altogether ruled out. Mainstream science would argue diprotodons, which in some respects resembled rhinoceroses without horns, may have died out 40,000 years ago.
www.mysteriousaustralia.*com*

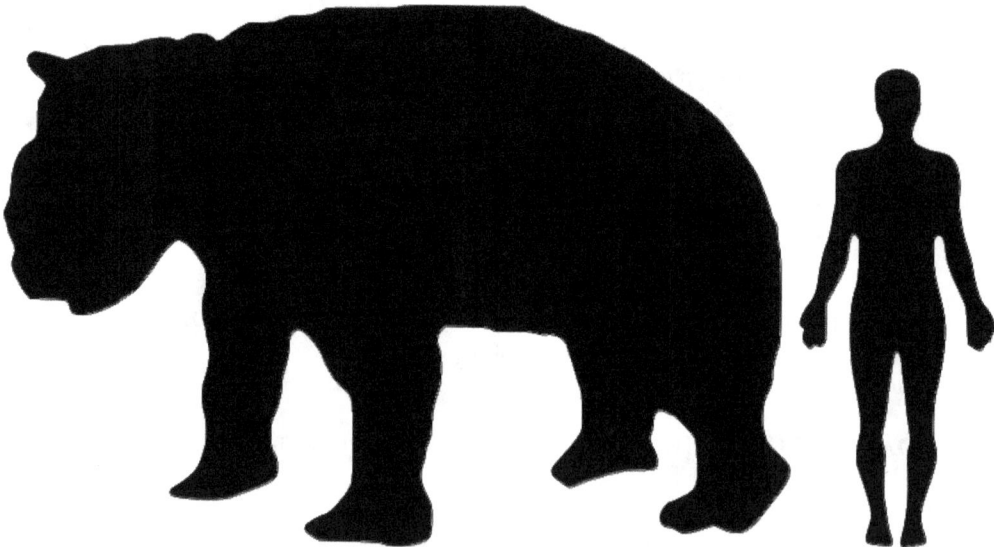

NEW YORK ANIMAL Strange animal caught in New York state in 1943. The animal's face resembled a fox's, the ears a wildcat's. The tail was short and stubby.
V: 398

NEW YORK CREATURE Two sisters saw this creature while driving on 15th March, 2008. The creature ran in front of the car. It was perhaps 4'-5' long. Its head may have been catlike. Its back was about 4' high, while the head was lower.
cryptzoology.com: March 25th, 2008

NEW YORK REPTILIAN A creature encountered in a basement in New York city. It had a reptilian head, human-like arms and legs, fangs and a long mouth. It was 6' tall and of a greenish grey colouration.

The man who encountered it was greatly shocked and his workmates, whose subsequent search failed to find it, thought at first he was having a heart attack.
www.naturalplane.blogspot.com: September 10th, 2010

NEW ZEALAND GREATER SHORT-TAILED BAT This animal (*Mystacina robusta*) may now be extinct. The last time one was observed was in 1967. [*]

[*] The **New Zealand greater short-tailed bat** (*Mystacina robusta*) was one of two species of New Zealand short-tailed bats, a family (Mystacinidae) unique to New Zealand. It lived on the North and South Islands in prehistoric times and historically lived on small islands near Stewart Island/Rakiura. Short-tailed bats were as adept at scrambling along the ground as they were at flying. This is the only known image.

NEW ZEALAND MOSASAUR   The mosasaur is supposedly long extinct, but there have been rumours of small mosasaurs off the New Zealand coast. New Zealand is one of the places where mosasaur fossils have turned up.
www.cfz.org.uk: 13th November, 2011

NGARAR   Legendary lizard of New Zealand. There are supposed to be two species. The larger has a crest on its back, toothed like a saw, and is 2-3' long. The smaller is about 18" in length.
nzcryptozoology.ucoz.com

NGOUBOU   An animal reported by the natives of Cameroon. These are said to be of two kinds, one river-dwelling, one land-dwelling.   Descriptions bear a resemblance to the triceratops.
EB: 84

NICOBAR SPARROWHAWK   This bird (*Accipiter butleri*) only found in the Nicobar Islands (see below) may be extinct, although it was possibly seen in the 1990s.

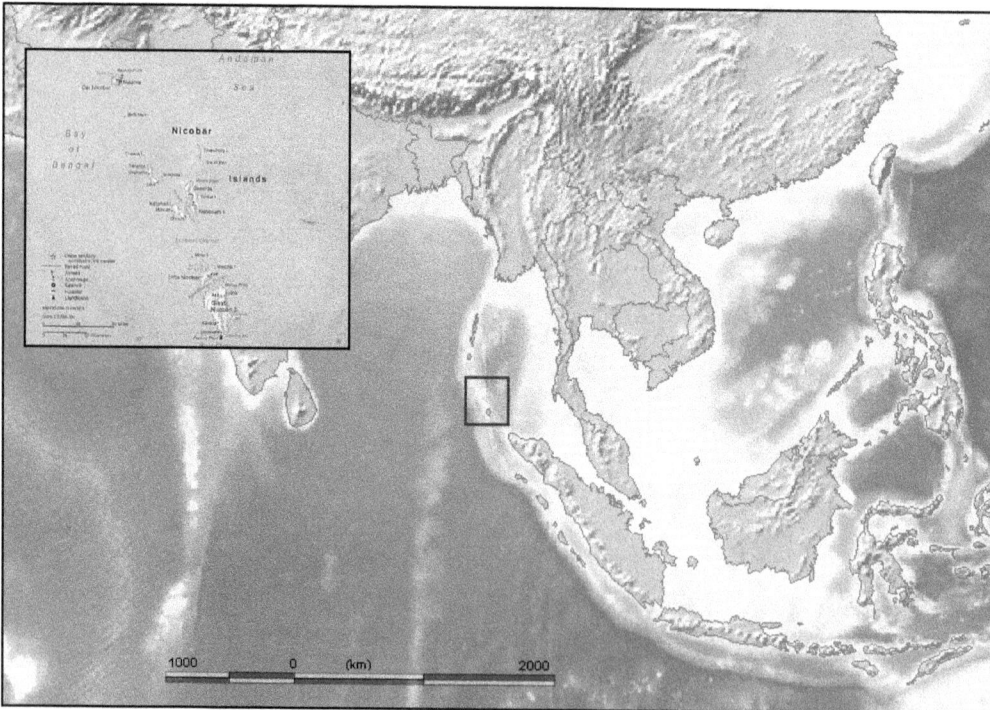

NIFOLOA   A kind of supernatural insect with a single tooth in Samoan folklore. The tooth is about 3" long. One nip will kill a human.
www.cfz.org.uk: 28th September, 2010

NIGHT-FLIER Term applied to a gigantic bat rumoured to be in Vietnam.
animaldiscovery.com/tv/lost-tapes

NINGEN A large Antarctic sea-monster resembling a human, perhaps an Internet hoax. It is supposed to have legs, arms, fingers, eyes, a mouth but no nose. Some reports give it a tail rather than legs.
forgetomore.com: 2007; *pink tentacle:* 2010

NIRIVILO A creature either half fox and half snake or a dog with a fox's tail in the legends of Chiloe (Chile).
LM

NOGGLE A creature having the appearance of a pony, also known as the *njuggle*, reportedly found in the Shetland Islands. If you mount it, it will dive into the water with you. It does this to millers, for it will stop a mill wheel with its teeth. The miller, going out to investigate, climbs on the noggle, to which he finds himself stuck and his watery doom is sealed.
NI: 37-38; www.shetlandtimes.co.uk

NON-HUMAN CREATURE This rather phantasmal being was seen in January, 2008, in Orlando (Florida). At Frontage Road, at the International Airport, it reportedly walked through a car. No description of the creature is available.
www.iraap.com

NORFOLK CREATURE A creature with grey/white long hair, quadrupedal and large-snouted. This was seen by a driver in Norfolk in 1986. It eventually stood on its hind legs and it was 6'-8' tall. A sighting of a similar creature took place in 2007.
www.paranormaldatabse.com

NORTH CAROLINA ANIMAL This was described as 'a monster of hideous mien'. It looked like a huge black bear with a lion's head and a wapiti's horns. It was the size of an ox, but longer. It was reported in 1873 and was seen by several people.
V: 438

NORTHICOTE FARM CREATURE This was seen by Steve Nicklin and a friend on a farm near Wolverhampton in 2004. It seemed to be over seven feet in height, was humanoid, had legs and arms and batlike wings. It leaped from one tree to another.
TSW: 82-83

NOVA SCOTIA HOMINID A large black hominid whose eyes gave out red light and which had a big red eye in its back was reported in the area of the Mule River in Nova Scotia in 1888.
zooform.blogspot.com: 18th October, 2011

NUIHI Legendary creature, perhaps in origin an elephant seal, reported from Easter Island.
ACC

NUKUPU'U This Hawaiian bird (*Hemignathus lucidus*) may now be extinct. It was last seen about 1996.

NUMSKE LEE KWALA A kind of sea-monster in the Pacific in the legends of the Comox Indians.

NUNG-GUAMA A beast of Chinese folklore with a taurine body, big floppy feet and clawed hands, not to mention a taste for human flesh.
karlshuker.blogspot.com: April 8th, 2011

NURUFILI In Araucanian belief, a monster that can catch you with its tail.
CR

NYUKUR A creature of Faroese lore, it looks like a horse and lures people onto its back, intending to drown them. It can extend the length of its back to accommodate the number of people wishing to ride it.
NI: 45-48

O

OCEAN RAT An unidentified creature washed up in 2011 at Crystal Beach, Galveston Island (Texas).
www.montauk-monster.com

OCONTO DRAGON This creature was observed on 27th October, 2007, by a group of people at Oconto (Wisconsin). It was flying overhead and was tan and white in colour. Its wings were like those of a bat and its tail was prominent. Later that night, smaller specimens were seen.
MW: 32-33

OFUGUGGI The reverse-finned trout, an Icelandic cryptid.
AZ: 291

OGSTON RESERVOIR CREATURE   The humps of a creature, perhaps even a monster, were observed in this reservoir in 2002. The skin of the animal was like that of an eel.
AZ: 155

OHIO ANIMAL   A strange creature seen by one G.G. Hibbard in Ohio in the 19th Century. It was the size of a small hound, built like a fox, coloured white with a black spot on its back. It had a bushy tail and its manner of progress resembled that of a cat. When Hibbard set his dogs on it, the creature knocked them to one side with its head.
BF: January, 2007

OHIO CATMAN   A large bipedal creature seen in Lorain (Ohio) in 1968. It approached the cabin of a family named Cataldo, first landing on the roof. The Cataldos saw a face of massive proportions looking in through the window. Then the creature ran off.

It was like a lion, but, when it ran, its gait resembled an ape's. Its handprints upon the window sill looked human, except they were reversed. Its footprints were also reversed.
I: 264

OHIO CREATURE   A large, brindled animal, the best part of 7' long from nose to tail and about 3' tall. It was killed by hunters in 1846 after a chase of about 150 miles and none could say what it was. A curious fact is that, during the hunt, it never crossed a fence.
V: 486

OHIO GIANT SNAKE   A snake, perhaps 100' in length, was reported in the Upper Sandusky in the 19th Century.
HR: 232

OHIO GIANT TURTLE   A mystery species of turtle, gigantic in size, was spied by B. Nunnelly and two of his brothers in the Ohio River (Kentucky) in 1983.
SKM: 42

OHIO PIRANHA   Piranhas have been reported from the Ohio river in Kentucky, indicating there may be an introduced population. A definite red-bellied piranha was captured there in 2007.
SKM: 115-6

OHIO WINGED CREATURE   As the witness was driving through Monroe County (Ohio) in November, 2008, in the early morning, he saw a large creature with wings folded about it in the middle of the road. It had glowing eyes.
www.iraap.org: rosales

OKEFENOKEE GIANT   A hominid 13' tall killed by a group of nine men, only four of whom survived the encounter. The story is found in a contemporary record.
TG: 74

OLATHE WILDMAN This creature, described as half man and half devil, was seen in 2007 near Olathe (Kansas), leading people to speculate that it was an escaped gorilla.
cryptomundo: July 22nd, 2007

OLD NED'S DEVIL A strange animal which attacked a man referred to as Old Ned in Smethwick, near Birmingham, England, in the 19th Century. Old Ned killed the beast which was then kept in a glass jar in the local Blue Gate pub in Rolfe Street. It has now disappeared and there is no description of the creature.
manbeast.blogspot.com

OLD SLIPPERYSKIN Gigantic creature which much affrighted the settlers of Vermont about the year 1800. It was said to be a huge bipedal bear, but, as it was also said to be able to fling stones, we cannot be certain, as a bear's paws are not designed for such activities. This creature frightened cattle and sheep, but does not appear to have preyed on them. It destroyed human constructions such as fences. Cryptozoologist Joseph Citro says it may have descendants active today.
VMG: 9-10

OLIVE IBIS (SAO TORME IBIS) A bird of São Tome whose continued existence is uncertain. The **São Tomé Ibis** (*Bostrychia bocagei*), also known as the **Dwarf Olive Ibis**, is a critically endangered bird that is endemic to São Tomé. Once thought to be a subspecies of the larger olive ibis, it is now classified as a distinct species.

OLOMA'O This Hawaiian bird (*Myadestes lanaiensis*) was last seen in 1980 and may now be extinct.

OMOK Sea-serpents in the legendry of the Asmat people of Irian Jaya, the western part of New Guinea.
FZ: July 17th, 2011

ONE-EYED APE This creature was a monkey rather than an ape, as it had a tail. Instead of two eyes, it had one eye in the middle. It was reported in Kentucky in 1831.
SKM: 18

ONTARIO ENTITY A correspondent of the website About.com (undated) claims he was living in the country when he heard noises outside and saw something red looking through the window in the front door. Looking out he saw an entity with what he took to be arms, waving them over its head. It seems to have been somewhat ill-defined. He obtained further sightings from an upper floor. The figure was of human shape. This occurred in 2006.

OÓKEPAM A monster that preyed on humans according to the Aonikenk Indians of Argentina. It has a protective shell.
PM

OOTSA LAKE MONSTER A head, resembling that of a large snake, was seen to emerge from this body of water in British Columbia. Then a second and a third appeared. The witness was Darlene Thompkins. The sighting is undated, but appears to have taken place in August, 2010. There had been a tradition of a monster in the lake.
Burns Lake Lake District News: August 23rd, 2010

OPHIOTAUROS In Greek myth, a creature which was like a bull at the front and a serpent behind. There was only one of its kind.

ORANG IKAN A kind of sea-dwelling humanoid with a mouth resembling a carp's. They are supposed to frequent waters near the Kai Islands of Indonesia. Japanese soldiers espied some in 1943 and Sergeant Taro Horiba inspected a dead one. It had red-brown hair which reached its shoulders, a short nose, a broad forehead and long webbed fingers and toes.
www.cryptomundo.com: July 17th, 2009

ORB OF THE VALLEY A lighted orb that seems to be alive and to show intelligence has been reported in the East Fairfield-Bakersfield area of Vermont. It responds to human speech and on one occasion actually accepted an invitation to enter a cabin. Matthew Williams once mentioned to existence of such creatures in England, while Rob Riggs, whose area of expertise is the Big Thicket area of Texas, has also mentioned such things.
VMG: 106

OSHADAGEA The Iroquois believed that this was the name of a huge eagle who caused rainfall.
www.probertencyclopaedia.com

OSLO FJORD SEA-SERPENT This had a horse-like head followed by three humps and was spied from a boat in 1933.
WW: 68-9

OSNABURG MINE ANIMALS In 1896 coal-miners near Osnaburg (Ohio) claimed there were mystery animals in the mines under the farmlands there. Only one creature had been seen, but another had been heard responding to its calls. It had a head the size of a large dog's and whiskers. It actually attacked one miner, but miners' lamps seemed to scare it. Footprints revealed six claws on each foot. It had emerged from the mines at times. Certainly, one was seen by miners. It bounded to a mine opening.
V: 495-6

OTTINE SWAMP MONSTER A creature known as "the Thing" is supposed to be on the prowl in this Texan marsh. Various tales speak of a huge creature or an invisible creature. The latter may have a tendency to shake cars or caravans.
MT: 99-100

OUAMPENANGOAG A monster reported from a number of Minnesota lakes. It may in fact be a huge sturgeon.
Wikipedia

OX-HEADED SERPENT Such animals feature in Greek folklore. They were reported in a flood in 1835 in Fonia and in the Rapsani Forest in 1891.
HG: 59

OZRUTA A wild man or giant living in the Tatra Mountains of Slovakia.

P

PACKDA A kind of ape mentioned in both Eberhart's and Newton's encyclopaedic works. It is supposed to exist on Palawan Island in the Philippines, but may be extinct.

PAINTED VULTURE This bird was supposed to be a native of Florida, but it may never have existed, but been due to misidentification.

PALE OCTOPUS An expedition undertaken by the universities of Oxford and Southampton, the British Antarctic Survey and the National Oceanographic Centre has discovered and photographed an octopus possibly unknown to science in southerly waters.
www.cryptomundo.com: January 5th, 2012

PANAMA CREATURE This mystery creature was allegedly killed by teenagers in Panama in 2009. It may have been an embryo. Its skin was leathery, its colour pinkish-grey, it had claws, a blunt nose and a thick tongue. Its belly was described as swollen. Identities suggested ranged from a shaved pit-bull terrier to an extraterrestrial.
www.associatedcontent.com: article 2209429

PARADISE PARROT This Australian parrot (*Psephotus pulcherrimus*) is supposedly extinct, last observed in 1927, but there is some evidence that it can still be found.
AZ: 70, 120

PAEKAKARIKI MONSTER A huge monster seen in the sea off New Zealand in 2001.
CFZY 2010: 174

PAWTUCKET CREATURE An animal which looked generally human but had a face resembling a wolf's, seen by four witnesses in a wood at this location in Rhode Island.
www.iraap.org

Psephotus pulcherrimus photographed alive in 1922 (Wikimedia Commons)

PECSAVARAD CREATURE On Mount Zago, in this region of Hungary, there were once supposed to live small dragons. They are described as looking like lizards or birds.
www.cryptozoology.com: December 19th, 2010

PENHUENCHE SNAKE A monster that is described as serpentine and humanoid. It is believed in by the Penhuenche Indians of Nequén, Argentina.
PM

PERM HUMANOID This was allegedly seen in Perm cathedral, (pictured on the left in 1916, courtesy Wikimedia Commons) Russia, in 1905. It was black, 3'-4' tall, its nose was like a beak and it had a wide mouth. It was inside a red aureole.

The witness, a 10-year old boy, was unharmed by the creature.
JHS: 1:3

PERTH PTEROSAUR A winged creature, at first taken for a bird, but then identified as a possible pterosaur, seen over Perth, Australia, in 1997.
www.ropen.com

PERUÈA LAKE MONSTER This lake in Croatia was created by the building of a dam in 1959. Surprisingly, reports are now coming in of a monster in its waters. It appears to have a long neck and a fat body has been described.
A&M 44: 22

PERUVIAN CREATURE An animal which seems to have been involved in mutilations of farm stock, reported from Puerto Huemy. It is the size of a dog, but looks more like a bear. Its chest is yellow, it has large fangs and it leaps over fences.
ufoinfo/humanoid: 2009

PHLACAL The elephant-goat, a beast in Armenian legend.
MAR: Vol. VII, 92

PHOENIX OF ANDORRA A huge bird reported by one Jesus Serras in 2001. It was of gigantic size and of a somewhat aquiline shape. It kept emitting a flash of light with red tones.
criptozoologia.blogspot.com: April 13th, 2011

PIATEK This looked like a griffin without wings. It was to be found in Armenian mythology.
Wikipedia

PICTON MONSTER A huge monster seen in the sea off New Zealand in 2001.
CFZY 2010: 174

PIG-FACED WOMEN Various stories of women born with the faces of pigs have circulated. They seem to have had their origin in England about 1638.

PINE HILL CREATURE An unidentified animal reported from Kentucky. It was described as larger than a cat, but smaller than a puma. It was said to have a row of spikes along its back.
SKM: 31

POGEYAN A big cat of uncertain species, grey, long-tailed, round-eared and rather large, seen by Sandesh Kadur near Mount Ananaudi in the Western Ghats, India.
scienceblogs.com

POLISH FLYING MAN This was seen in the north of Poland in 2000. Two young men saw a humanoid flying above the treetops. The figure was estimated as two metres tall. It had long hair.
purpleslinky.com: September 6th, 2009

POLLO MALIGNO In Colombian folklore, a gigantic chicken which will eat humans. (I wonder if it thinks, "They taste just like chicken"?).
Wikipedia

POMBERO A creature believed in by the Guarani Indians of South America. It is a humanoid and, apart from playing tricks on farmers, it makes women pregnant, sometimes by just touching their hands. It is said to live in the jungle. Clearly mythologised, there may yet be some creature behind the myth.
Wikipedia

POOL CREATURE A creature seen by Norman Dodd in a pool in Cannock Chase in 1976. It appeared to be sunbathing. Its head was like a snake's. Its skin was oily. It had flippers or small feet near the front. Mr Dodd did not see the back end.
TSW: 42

PORCINE ANIMAL This strange beast had the general aspect of a pig, but was larger and hairier. It drove off two dogs that were set on it. It appeared in Roscommon (Ireland) in 1898.
MAI: 87

POTEET CREATURE Off the Poteet Highway (Texas), this animal that looked like a dog came scratching at the windows of a house and then flew onto the roof. Police are said to have turned up and driven it off with gunfire.
MT: 98

POTOMAC RIVER MONSTER A possible river monster was spotted and photographed

on the Potomac on September 14th, 2010
www.cfz.org.uk: September 29th, 2010

POWYS FLYING CREATURE In March, 2001, several persons including a naturalist in this part of Wales saw a volant animal about 2.5' in length. No wings were observed: it proceeded by undulation. Its head was like a sea-horse's, its colour green and it had a long tail with flukes.
www.iraap.org: rosales

PRIMEHOOK SWAMP CREATURE This was about 2.5-3' tall, its face was like a pug's and it was endowed with a long tail. It was seen by a driver in Brickhill Beach (Delaware) and there have been other reports.
About.com: paranormal phenomena

PROVIDENCE BIRD A bird the size of a human with wings, spotted in Tierra del Fuego. It was seen in 2000 and blamed for killing sheep.
PM

PTAK OHNIVAK Name used for the firebird in Czech folklore. There is an article on the firebird itself in the author's *Dictionary of Cryptozoology* (2004).

PUEBLA CHUPACABRAS Goatherds fear such a beast is in the vicinity of Puebla, Mexico, due to attacks on goats in the area.
www.ghosttheory.com: September 14th, 2010

PUERTO RICO GARGOYLE This creature has been reported for many years, chiefly from Guanica. It is said to have bat-like wings and some believe it to be a bird. It has on at least one occasion supposedly attacked a human.
inexplicata.blogspot: August, 2010

PUHKIS A kind of dragon in Latvian mythology.

PUMPKINTOWN CREATURE A strange white animal, larger than a dog, perhaps with four eyes, attacked a car at this West Virginia locale.
WT: 24-25

Q

QUANG NGAI BEAST A mystery animal of Binh Son District, Vietnam, whose existence has been inferred from paw prints and roaring.
livingdinos.com: July 17th, 2011

QUEENS BEAST An animal seen in Queens (NY) by a traveller and his girlfriend. He saw

The firebird of central European folklore

the animal, which ran on four legs, on 100th Street. He reckoned the animal was about 7' tall. The girlfriend said it was too big to be a horse.
cryptoflorid.ning.com: April 5th, 2010

R

RABISHA WATER BULL The monster of Lake Rabisha, Bulgaria, is supposed to have a bull's head, a strong male body and a fish's tail. It is said that, in the old days, each year a beautiful young girl was sacrificed to the monster. This custom was eventually stopped by the water bull himself, who was so smitten with one of the girls that he persuaded his sister, who had magical powers, to make her immortal and no more sacrifices were necessary.

The presence of wels catfish in the lake may have fuelled the legend.
www.novinite.com: March 5th, 2010

RACER BEAR Alternative name for the Ranger Bear.

RAMIDREJU In the north of Spain, this legendary animal, though having fur, looks much like a snake. This fur has a greenish tinge. One is supposedly born every hundred years to a weasel or a marten and its skin is said to have curative properties.
Wikipedia

RANGER BEAR A large and very fierce bear said to dwell in the United States. On its breast it has a white star or crescent, as has the Asiatic black bear.
V: 120

RAPUWAI A hominid in which the Maoris of New Zealand believe. Its hands are very strong and they can strangle humans. They are said to be slow and clumsy. This creature has been compared with *Homo erectus*, thought to be the hominid which preceded modern humans and out of which they evolved.
CFZY 2008: 164, 166

RAROG A whirlwind spirit, variously conceived as being in the shape of a hawk or a dragon, found in Polish folklore.
Polish Wikipedia

RATTLESNAKE ISLAND SNAKE This peculiar animal, which does not appear to have been a rattlesnake, was supposed to live on the island of this name in Lake Erie. It breathed a breath which, if you inhaled it, would kill you.
MLBOB: 45

RAUDKEMBINGUR A cryptid whale of Icelandic lore. It seems to be much the same as the hrosshvalur (q.v.) except that it is brown instead of grey.
AZ: 291; *WW:* 37

RAUMATI MONSTER A comparatively small monster was seen in the sea off this New Zealand location in 2007.*CFZY 2012:* 174

RED BEAST A red animal resembling a kangaroo was seen at Ivanovo, Ukraine, on 7th August, 2009. There had recently been animal mutilations in the area.
HSR: 2009

RED-EYED WOLF-BEAR A black creature, larger than the American black bear, with a bushy tail and pointed snout. It was at first quadrupedal, but, when it stood upright, one could see that its front legs were shorter than its hind legs. It was witnessed by a young couple in 1980, but this sighting was not the first.
MAWV: 84

RED MACAW   Unidentified bird (left) painted by Roelent Savery (16th Century) in the vicinity of a dodo, which forms the centrepiece of his painting. This bird is unidentifiable, but perhaps flourished on Mauritius, home of the dodo.
Shukernature: March 24th, 2011

RENGO SNOW MAN   This was reportedly seen in Chile in 1956-7. It was about 6.5' tall and supposedly clad in furs.
PM

RIFT VALLEY CREATURE   A large carnivore with spiny fins along its back has occasionally been reported from the Great Rift Valley in Kenya. Descriptions resemble the dimetrodon, an extinct lizard with a sail on its back. It had mammalian rather than reptilian dentition.
www.americanmonsters.com: May 31st, 2010

RIO CUBUY CREATURE   A humanoid, perhaps 7' tall, with brown skin, seen in Puerto Rico in August, 2008. It had a pointed head and three claws on each hand. It appeared to

stick out at the shoulders. The creature ran off.
www.iraap.org: Rosales

RIO GRANDE CREATURE A bipedal creature, about 4' tall, with black and grey hair, seen in Puerto Rico in August, 2008.
www.iraap.org: rosales

RIVER-WOMEN Two creatures in the mythology of the Shilluk of Sudan. They were women to the waist, thereafter crocodiles. Their father, Ud Diljil, was crocodile on his right side, human on his left.
www.answers.com

ROCK APE Unidentified ape with a penchant for returning fire when rocks are thrown in its direction, reported by Americans serving in Vietnam during the war. No description is to hand.
texascryptidhunter.blogspot.com: September 6th, 2011

ROCKHAMPTON CREATURE An animal seen but not identified by a teacher in this Australian city in 2011. Its fur was thickish and brown with rings on the tail which did not contrast greatly with that of the fur as a ring-tailed lemur's would.
The Morning Bulletin: April 8th, 2011

ROCKY BRANCH TROLL It had a wide forehead with white hair, together with blue eyes. Its trunk was large, but its arms were thin. It was a humanoid figure.
HSR: 2009

ROOK-HEN HYBRID It was believed in Clare, Ireland, that such a hybrid could sometimes occur.
MAI: 98

RORE-TROLD A shape shifting creature said to inhabit the Rorevand, a lake in Norway. It can assume the form of a horse or serpent or even a load of hay.
NM: 243

ROSEMONT CREATURE An animal described as a grey, hairy thing has been sighted near this West Virginia locale.
zooform.blogspot.com: 2008

ROSWELL RAT It has been said that, after the Roswell Incident in 1947, some materials were flown to Carswell Air Force Base (Texas). It has further been said that the material contained large animals resembling rats. They moved as a group, were very intelligent and very strange.
www.monsterusa.com: November 21st, 2010

RUAEO Giant hominids of Maori legend which could reach a height of 3m. They made

stone tools and buried their dead.
www.mysteriousaustralia.com

RUBBER-FACED BEAR A huge bear in the lore of the Indians of British Columbia. It has no hair on the top of its head or its face. Its nose is short. G. McIsaac is inclined to identify it with the huge short-faced bear (*Arctodus*) which received wisdom says became extinct in Canada about 12,500 years ago.
BH: 59-62

RUFFORD LIZARD A green and red lizard the size of a large dog has been reported in this area of Lancashire.
www.paranormaldatabase.com

RUNAWAY BAY CREATURE A dead creature discovered at Runaway Bay (Texas), which seems impossible to identify. It was found on the golf course at Hole 14.
www.startelegram.com: 18th January, 2010

RYE CREATURE A couple were walking near Rye (Sussex), when a creature that looked like (but presumably wasn't) a horse ran past. It jumped a fence and was heard to land in a deep pool on the other side. We are not told anything of the dimensions of the pool, so we do not know if this was a water-dwelling creature, reminiscent of the kelpie of Scotland.
K: 313

RYUGYO A mystery fish reported from Japan. It is said to resemble a sturgeon.
Wikipedia

S

SAAPAIM Sheep-sized shaggy animal with powerful claws, reported from Tierra del Fuego.
PM

SAENEYTI In Icelandic lore, these are cattle that live in the sea, but come ashore to mix with their terrestrial counterparts.
AZ: 291

ST CHARLES SERPENT Mrs J. Bishop of this Missouri location reported a spotted flying snake in 1911.
HR: 246

ST IVES CREATURE A witness described having been chased at the beach at Hayle Towans near St Ives (Cornwall). He never saw his pursuer. He made it to a beach house and heard grunting from outside.
About,com: paranormal phenomena

SALT LAKE CITY ANIMAL On August 8th, 2010, a creature on all fours with pale skin and arms as long as its legs was espied in Salt Lake City. The face could not be discerned and the head was an unusual shape. The body appeared hairless. The skin was somewhat transparent, allowing the witness to see ribs and spine.
JHS: 1:5

SAMOAN WOOD RAIL (left) This bird's continued existence is in doubt, though it was possibly seen as late as the 20th Century.
Wikipedia

SAN ANTONIO CREATURE This was espied by F. Ramirez of San Antonio (Texas) on the roof of his garage. It appears to have been seen after dark.

It was large with giant wings. The face of the creature was humanoid, though elongated, and it extended into a long beak. Although the wings were cloaklike, Ramirez heard them flapping as he departed at some speed.
MT: 12

SAN BENITO FLYING CREATURE A creature resembling a bird which was rumoured to exist in this Mexican town. It could be, not a bird, but some kind of bat.
CR: 2:1

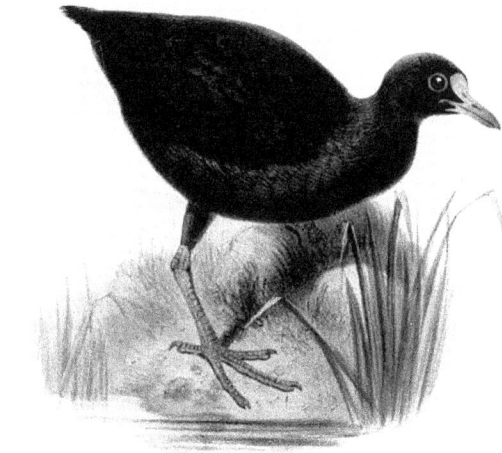

SAN DIEGO CREATURE In June, 1890, a strange creature with vespertilian wings was seen over San Diego (California) by a boy named Diller and a man named Marvin. The latter saw it over Switzer Canyon.
cryptozoology.com

SANDWALKER If you are in the Arabian deserts, make sure your camel is well-guarded by night, lest the legendary sandwalker eat it. This legendary beast is the size of a horse, but has a beak, a crablike body and a scorpion's tail.
Wikipedia

SANDY HUFF BEAST Unidentified animal with canine characteristics reported from West Virginia, proceeding sometimes on two legs, sometimes on four.
MAWV: 90

SÃO PAOLO CREATURE A strange creature seen outside twice by a witness and once by his sister in 2009 at São Jose de Rio Preto. It was described as grey with yellow stripes on its back.
ufoinfo/hominids: 2009

SÃO SEPE HUMANOID A bipedal creature looking like a big dog which attacked a 20-year old at this location in Brazil in 2009.
ufoinfo.com/humanoid: 2009

SARIMANOK A bird in the mythology of the Maranao of the Philippines. Catch one and it will bring you good luck.
Wikipilipinas

SCALAWAG Creature perhaps identical with Bigfoot reported to live in the Blue Ridge Mountains of North Carolina. A witness described it as having orangey, sparse hair and an unpleasant smell.
About.com: undated

SCALLOWAY CREATURE A creature resembling an upturned boat, seen offshore in 1810 from Scalloway, Shetland Islands. It seems to have had a sort of ridge on its back. It lingered in the area for a fortnight.
NI: 166-7

SCOLOPENDRA A kind of huge sea-monster in ancient Greek belief. They had hair on their heads, feet along their sides for propulsion and a crayfish-like tail. It gave its name to a well-known genus of centipede.
www.theoi.com

SCOTLAND DEVIL A vague term used in the United States for unidentified felids.

SCULTONE A kind of dragon in Sardinian tradition.
Italian Wikipedia

SEA BATMAN A creature looking like the Batman symbol of the DC Comics hero has been seen on a volcanic ridge, off Vancouver Island, Canada.
metronews.ca: May 27th, 2012

SEA BULL A creature of Armenian folklore, whose existence has been inferred solely from the noise it allegedly makes.
MAR: Vol. VII, 92

SEA PET Name given to a kind of sea-serpent witnessed by J. Barker off Washington state in 1884.
SM: 35

SEA-RAM   A black creature with two white horns and appalling halitosis said to dwell in Norwegian waters.
WW: 70

SEATTLE SABRETOOTH   A partially transparent sabretooth seen by a woman named Jenny Burns when walking her dog. The paranormal element here requires careful study, but should not be dismissed.
The 'C' Influence: August 29th, 2010

SEGUIN TRILOBITE   An animalcule resembling the supposedly extinct trilobite (pictured below by Heinrich Harder) was said to have been seen at this Texas location. It scuttled back into the earth.
MT: 91

SELAMODR   A huge kind of seal, which, in the lore of the Icelanders, protects other seals of a more commonplace size.
AZ: 291

SERENGETI ENIGMA   An unidentified animal seen in the Serengeti National Park, Tanzania. It may be either an unknown animal or a mutant. It has the body of a gazelle, the head resembles a camel's and it sports a leonine mane.
www.ippmedia.com: December 20th, 2010

SERPENT OF DEAD CREEK   A creature which attacked three fishermen of Vermont in 1909. The head was black and hairy, the throat had large scales and the animal's

circumference was that of a sugar barrel.
VMG: **88**

SEWER CREATURE In the Sussex Water Treatment and Water Management Plant, there is supposed to be a strange shuffling figure that stops when lights are shone on it. The sound of voices may indicate there is more than one of these creatures and that they can converse.
P 46: 13

SEWER PIGS There was a tradition in Hampstead, London, in 1851, that feral pigs were on the rampage in London sewers. They were descended from a gravid sow that had become trapped in a sewer. If so, the question arises as to whether their descendants still rove there.
naturalplane.blogspot.com: December, 2010

SEWERFACE A humanoid creature allegedly photographed in a sewer under Newtonbreda, Northern Ireland, by CCTV.

SHACHIHOKO In Japanese lore, an animal with the head of a tiger and the tail of a carp.
www.americanmonsters.com: October 28th, 2010

SHARPSBURG ANIMAL Unidentified animal seen in Sharpsburg (Pa.) on June 18th, 2011. It was about 2' tall and 2.5' long. It cantered like a horse. It seemed hairless or short-haired. Its snout and ears were pointed. It had a tail. It was slim, but did not seem undernourished.
Stan Gordon

SHAWSVILLE CREATURE In this area of Virginia in 2002 reports started coming in of an unusual beast. It had light hair, was bipedal and was about 6.5' tall.
MAV: 77

SHEEP MUTILATOR In Navaho belief these are diminutive beings with faces like those of apes.
AE: 122

SHELBURNE BIRD A strange bird which the observer could not identify, seen in her back garden in Shelburne (Vermont). Its wingspan was 8'-10'. The bird seems to have had a somewhat menacing look.
VMG: 103

SHEPLEY LION A possible lion and possible lion-cub were seen on the railway line near Shepley (Yorks) on November 6th, 2011.
paranormaldatabase.com

SHI-JOU In Chinese lore, a beast supposed to look like an ox's liver, but it differs in that it has two eyes. You can eat it, but it will grow once more, one hopes not inside you.
CMS: 237

SHON(E)Y A monster supposedly found in the North Sea, between Britain and Denmark. The medieval Norse believed in it. It was said to have a coral cave beneath the sea, in which it kept prisoners, before it ate them.

It was supposed to have left the area, but to have returned in the 12th Century. This was inferred from the large number of bodies that turned up on Holy Island.

The Norse also frequented the Western Isles of Scotland, where Shoney was revered and where the origin of belief in him perhaps lay. In Gaelic he was called *Seonaidh.* On the Isle of Lewis it was the custom every year to offer him a pot of ale to ensure good fishing. This custom continued until about 1660.

It is possible Shoney was originally a sea-divinity, his name applied to sea-monsters seen off the coast, hence his being thought to dwell off the North Sea. A report in the *Shields Gazette* of 1946 dealt with a huge head and neck seen in that sea from a ship called the *Black Eagle.* Some of the crew entered a motor-boat and pursued it, but did not catch up with it. Monsters of this kind would have been identified with Shoney.
N&T: 175-196; *MTBM* 28-29

SHORTNOSE CISCO A fish (*Coregonus reighardi*) that inhabited the Great Lakes of North America, now possibly, but not certainly, extinct.

SHUANG-SHUANG An animal in Chinese lore. It has a number of shapes.
CMS: 247

SHU-MAO LIU-SHO Bird of Chinese lore. It sports six heads, very useful if five are cut off.
CMS: 256

SIBERIAN CHIPMUNK   An animal (*Tamias sibericus*) found in northern Asia. It has recently invaded a number of countries to which it is not native.

The invasions seem to be due to escaped pets and their offspring. They breed rapidly, but can carry ticks harbouring Lyme disease.

There is a considerable number in the Paris area and some have come to Britain through Eurotunnel. There has also been an escape in Britain.

In Ireland, they have been spotted in Co Waterford. Some are also to be found in Denmark, not far from Copenhagen.
www.telegraph.co.uk: 14th November, 2011;

www.thesun.uk: July 30th, 2008; invasivespeciesireland.com; cryptodane.blogspot.com

SIGBIN A blood-sucking creature of Filipino tradition. It looks like a goat without horns. It stinks highly. It has big ears and a tail it can use as a whip. It can make itself invisible. Wikipedia

SILVER LAKE MONSTER This creature was reported in the 19th Century from this lake near Perry (NY). It was said to be 10'-12' in length and dark, with a dorsal fin. Ersjdamo's Blog: August 12th, 2012.

SIMIOT A kind of primate, somewhat like the hybrid of an ape and a monkey, in Catalan tradition. cryptologos.blogspot.com: April 3rd, 2011

SIYOKOY These are tentacled creatures of the sea with an appetite for humans. They are sometimes depicted with webbed hands and feet. Wikipilipinas

SKIPWITH BEAR A bear was reported in this area of Yorkshire in 1975. It was dark brown. www.paranormaldatabase.com

SLIBINAS A dragon of Lithuanian legend, usually polycephalous.

SLIEVE RUSSELL CREATURE An animal discovered in Cavan, Ireland, in 1896. It was nearly as big as a donkey with grey hair, a pig-like head, small eyes and a bushy tail. It defied identification. *MAI:* 87

SLIGO MONSTER An Irish legendary creature, half bear and half dragon. *MAI:* 74

SMALL BIGFOOT These are believed to exist in various parts of the United States. *AE:* 43

SMETHWICK ANIMAL *see* **Old Ned's Devil**

SMOKE-COLOURED CAT A large animal seen in Victoria (Australia) in 1985. Its tail was curved. *ABC:* 314

SMOKEMAN In Essex on October 14th, 2010, a motorist saw a human form, apparently made of brown smoke, materialise in nearby railings. It looked to her like a man-beast with a small head. Then it disappeared, apparently sucked back into the railing. www.cfz.org.uk: December 20th, 2010

SNAKE KING In the folklore of the Netherlands, this is a snake bigger and fatter than other snakes, coloured white or yellow. In Gröningen it is called the *woaterslang* or *otterslang,* in Friesland the *kroentsjeslang.* In Drenthe it is said there is one king snake for every thousand snakes.
See also **King Snake.**
www.cfz.org.uk: 18th November, 2010

SNAKE-HEADED DOG It is quite possible that this beast is no dog at all. One cannot, in fact, tell what it is. It was seen by Sheila Charles and her son Shane. It crossed the road in front of their car at Magalia (California) in 1996. Its head looked like a snake's, it had a long, thin neck (perhaps 30" long), short front legs, somewhat longer back legs, black and shaggy fur and no tail. Another motorist confirmed the sighting.
AZ: 35

SNAKE-MAID There were two of these who lived in an unspecified lake or sea in Algonquian legend. They were snakes below the waist. Originally human, the change had taken place when they were swimming.
Leland: 270-71

SNOW PLOW ANIMAL This is an animal said to be able to plough its way through the snow. Alfred L. Danity claimed to have trailed it. He felt that a white animal with canine features and the body of a small bear on whose dead body he had come across in 1899 might have been the same kind of animal.
V: 130-1

SOLOMON ISLANDS GIANTS Gigantic hominids are reputed to live on the Solomon Islands. M. Boirayon asserts there are hundreds and perhaps thousands of them on Guadalcanal. There are two types, a tall one that can reach 10' with black, brown or reddish hair, red eyeballs, a flat nose and wide mouth and a smaller one like wildmen who are still taller than humans. Human-giant hybrids have occurred and there are humans of giant descent living in Tangarare. In the past they were said to eat humans, but a leader called Luti Mikode is supposed to have put a stop to this in the 20th Century. This Luti Makode is said to have met an Anglican bishop in 2000.

There is supposed to be a lighting system in their caves and they are credited with a form of writing. He asserts they are also to be found on Malaita, Santa Isabel and Choiseul. On Guadalcanal they are called *Mu-mu,* on Malaita *Ramo.* The Ramo are said to use a kind of wooden sword called a subi.
SIM: passim

SORTEDAMSSOEN MONSTER At a lake in Copenhagen people claimed to have observed monsters in the 1970s and 1980s. Earlier, in 1928, a couple of porpoises appeared in the lake. How they got there is a mystery, as is how they finally disappeared.
WW: 50-51

SOUTH CAROLINA SERPENT This flying serpent was allegedly seen by three sisters of Darlington County in 1888, nor were they the only witnesses. It seems to have attacked a weathercock on a church in Grassland later in the day.
HR: 236-7

SOUTH CAROLINA THUNDERBIRD It had white and blue feathers and a small beak. It had a plume on its head. Its wingspan was 12'-15'. It was seen by a motorist in 2009 on his way to Bamberg (SC). A similar but smaller bird - wingspan about 8' - was seen in 2010 by the same motorist on the way to Denmark (SC).
About.com: Paranormal Phenomena

SPALDWICK CREATURE A strange parasite, said to have entered a horse and killed it in 1586 at Spaldwick (Cambs). It is described as having had tentacles. It escaped from the horse and was killed with a dagger.
P 52: 73

SPANISH CROCODILE In 1998 this creature was said to be lurking in Juan Carlos I Park in Madrid. Some opined it might be an alligator. Searches proved fruitless.
criptozoologos.blogspot.com

SPANISH LION In September, 1517, a lion was reported on the recently flooded streets of Valencia, Spain. For some reason it was called the Lion of Germania. Some said the creature was no lion, but a roaring ox.
criptozoologos.blogspot.com: September 27th, 2011

SPECKLED JAGUAR Jaguar-sized felid covered with black speckles reported from Peru. The speckles are solid, unlike the rosettes of true jaguars.
shukernature: May 11th, 2009

SPECTRE MOOSE OF MAINE An exceptionally large moose, said to have been roaming Maine for at least twice the length of the average life-span of a moose. It was reported in 1891 from Lobster Lake. Although fired upon, it was apparently uninjured. Reports continued until the 1930s. Although it seems to have been brown, a white bull moose seems to have become confused with it at some stage. It may be that it was originally whitish rather than white and it is perhaps identical with the King Moose or Ghost Moose of Vermont.
cryptomundo: February 18th, 2007; *VMG:* 12

SPLITBACK DEMON A lizard-man or reptoid reported from Northumbria. It is about 5' tall, with scaly skin, a vivid red mane, a spinal fin, a long tail, talons 4" in length and a snout like a dog's.
forteanzoology.blogspot.com: May, 2009

SPOTTED ANIMAL An animal the size of a German shepherd with spots and a long tail was seen in the vicinity of Little Orleans, Maryland, in 1957. There were several witnesses.
V: 314

SPOTTED CREATURE This was observed by Australian Air Force personnel in 2005 near Weipa on Cape York peninsula. They estimated that it was as tall as a man's waist. It bore some resemblance to an hyena. Subsequent suggestions that it was a speckled boar have been discounted by the witnesses.
ABC: 233-4

SPOTTED WATER DOG These animals, in the belief of the Shasta Indians, were supposed to drown people in the water. The corpses accrued spots like their killers'.
CR: 2:1

SPRING LAKE GIANT SNAKE A huge snake reported in the water from this part of New Jersey in 1895. It had a face with alligator characteristics.
MNJ: 90

SQUASC An animal in the folklore of eastern Lombardy in the north of Italy. It has a human face and a squirrel-like body, but no tail.
Wikipedia

STEARNS BAYOU WATER MONSTER Stearns Bayou is in Ottawa County, Michigan. Here, in 1909, a monstrous creature was heard crossing the land before it entered the water. Its head resembled that of an hippopotamus, its body a soft-shelled turtle and it had a tail estimated at 10' in length. It was seen by a number of witnesses in 1909 and later in the same year by a man called McCabe.
www.michigansotherside.com

STELLER'S SEA BEAR A bear-like creature, white in colour, reported to G.W. Steller (1709-46) in the Kuril Islands. It growls like a bear. Steller was cautious about the existence of the beast, as he could discover no witnesses.
karlshuker.blogspot.com: January 4th, 2011.

STELLER'S SEA COW This marine animal (*Hydramalis gigas stelleri*) has supposedly been extinct since 1786. However, there have been a number of possible sightings in northerly waters.

In 2010 angler Chuck Corby saw a huge animal in the Pacific off Washington state and this

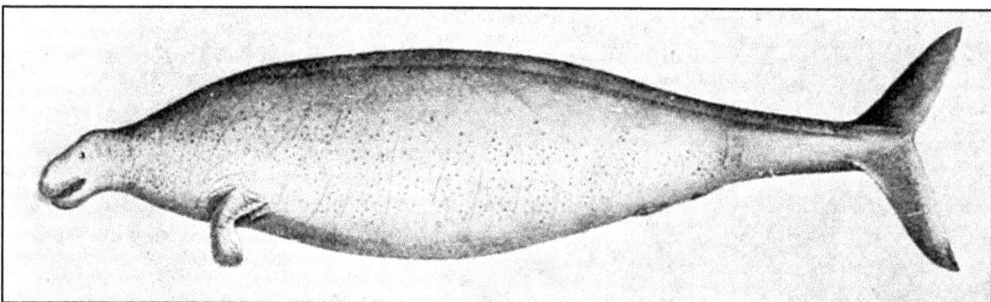

conformed in a number of respects to the sea cow. It is possible that the range of this animal is wider than supposed, accounting for reports of sea monsters in Hudson Bay and Baffin Bay, Canada.

www.cryptomundo.com: September 23, 2010; *ACC*

STERLING BEAST　This beast was seen in Whiteside County (Illinois) in 1938. Its background colour was tan with yellow spots in the foreground. It had white shoulders. It was larger than a mastiff.

V: 233

STICK FIGURES　Three black humanoids, thin like sticks, were seen walking through the snow into some trees in Wayne County (Pa) in 2009.

ufoinfo.com/humanoid: 2009

STRANGE ANIMAL　This felid was so described in the *London Times* of 1908. It resembled a lion somewhat, but had no mane. It was tawny with a reddish tinge. It appeared to have black spots and a black spinal stripe. There were dark markings like chain links from the tail to the centre of the back.

MAL: 164

STRANGE CREATURE　A man holidaying with his family in the Rocky Mountains in Canada went out one night to fix the independent generator. He beheld a human-like creature. It had no eyes, only indentations, though the light of his torch seemed to affect it. It was about 6' tall and you could see its organs through its skin. It looked as though the skin were too tight. The witness returned to the sanctuary of the cabin.

About.com: Paranormal Phenomena

STRANGE FISH This was a dark brown creature, looking like a cross between a fish and a sea-lion, which was brought to Bristol in 1763. Identification was not possible.
MSB: 167

STRIPED MYSTERY WILDCAT This animal is described as being larger than a lynx or ordinary wildcat. It was striped and captured at Tonopah (California) in 1925.
V: 160

STRIPED PONY A breed of pony striped yellow and white has been reported from New Guinea.
AZ: 74

STRIPED WOLF The *Hong Kong Telegraph* (January 30th, 1937) reports the killing of a wolf which may have given rise to rumours of a beast called the Kowloon Tiger. This implies the wolf was striped. It may actually have been an hyena.
www.cfz.org.uk: November 3rd, 2010

SUAN-YU A bird in Chinese lore.
CMS: 259

SUDANESE PTERODACTYL In 1988 a boy in Sudan reported a creature 4'-5' tall on a roof. It was both feathered and leathery, with a projecting bone at the back of its head and a long, leonine tail. The source, unfortunately, does not say where in Sudan this is supposed to have occurred.
modernpterosaurs.blogspot.com: June 2nd, 2010

SUILEACH In Irish legend, this was a sea-monster with many eyes, said to frequent Lough Swilly, an inlet of the sea in Donegal.
MAI: 78

SUMMIT LAKE WATER DOG Water dogs with long ears and light brown coats with white spots were said to have emerged from this Nevada lake to eat fish guts, according to the Paiute Indians.
www.strangeark.com

SUMTER ANIMAL At this Georgia location a policeman encountered a strange animal in 1882. It may have been an hyena, but he also suggested other possible identities.

SUNDERBANS CAT A small kind of felid with a long tail, brown in colour with black spots or patches, caught by a camera trap in the Sunderbans Tiger Reserve in India in 2012. It is smaller than a leopard but larger than a wildcat. It may constitute a new species.
www.cfz.org.uk: March 22nd, 2012.

SWIMMING LOG A creature reported from sundry Finnish lakes. It resembles a log, but, when it lifts its head, this is revealed to look like a salmon. Some swimming logs are white.

They are 3m-4m in length.
WW: 97-99

T

TAHITIAN GOOSE This bird was reported in times gone by, but nobody today knows what it was. Perhaps it yet flourishes, called by some other name in modern times.
Wikipedia

TAKITARO A giant fish said to be found in a lake in Japan, capable of attaining a length of 10'. One may have been captured and eaten in 1917. A couple of sightings were reported in the 1980s. Sonar investigation has shown there are large fish in the lake, Lake Otoriike.
www.monstropedia.org

TAKLA LAKE MONSTER There is said to be a monster in this Canadian lake. One description makes it resemble a crocodile. G. McIsaac thinks it is a plesiosaur.
BH: 43-46

TALONED TAPIR A species of tapir so called. It has been reported in a number of places in South America.
PM

TAMBARAN Giants, black and hairy, 15' tall, reported from Papua-New Guinea. A police patrol was supposed to have encountered a group of about twenty in West Sepik Province in 2002.
SIM: 29-30

TANKONGH A creature rumoured to exist in Guinea, compared by locals to a small zebra. It is also said to have small tusks.
DSC: 166-7

TAPETYWASON Beasts bigger than horses with large ears and bovine tails. They were reportedly found at the Straits of Magellan in 1591.
PM

TARMFISK Described by a Swedish traveller in Pennsylvania in the 17th Century, this fish looked like a rope a quarter of a yard long. It had four little bowels at each of its corners. It would suck food in through two of them and eject them from the other two.
MAP: 27

TATRA CAVE DOGS Mystery dogs said to inhabit caves in Poland. It has been suggested they live in caves by day and rove the mountains by night. They may be a feral population of domestic dogs.
BR: 17

TAUPO MONSTER A huge creature seen off the coast of New Zealand in 1990.
CFZY 2012: 174

TE-WHEKE-A-MUTURANGI Kupe, one of the legendary heroes of the Maori, defeated this creature, which was a huge octopus. One opinion is that the creature is symbolic.
Wikipedia

TEMUKA MONSTER Creature reported in 1974 off the mouth of the Oran River, New Zealand.
CFZY 2012: 174

TENATEE Giants of Kaska Indian lore, said to dwell in holes in the ground.
TG: 93

TENNESSEE ANIMAL An animal that was killing pigs in the vicinity of Clarksville in 1892. Hair of the animal revealed the skin was white, the tips reddish.
V: 600

TERMON ANIMAL In 2008 a woman in Termon, Ireland, encountered this creature in her cowshed. It was dark (black or brown), had a long nose, a hump, four legs and a long tail. It had a juvenile creature with it. Near the same location an unidentified creature blocking the road was seen in 2005 by one Cyril Lavelle. It was dark coloured with a large head. It

plunged through the roadside fence.
MAI: 65-66

THANET FISH   This creature was said to have cast itself ashore on the island of Fishness, Thanet, on July 9th, 1574.   It was 66' in length and its lower jaw opened to 12'.   It was possible for a man to stand upright in the socket of its eye.   This distance between its eyes measured 12', while it measured 14' from back to belly.   The tail was of the same thickness. Three men could stand upright in its mouth and a man might have crept into its nostrils, though, had I been there, this is an opportunity I would have foregone.   This account is one which will no doubt attract suspicion.
www.kentmonsters.blogspot.com

THEVRUMINES   The Etruscan name for the Minotaur (article in *Dictionary of Cryptozoology*).

THORNVILLE CREATURE   In Thornville (Ohio) at 4.30 a.m. on June 12th, 2009, a witness saw from her office a humanoid creature appear, seemingly out of nowhere, in her garden. It walked a few steps and disappeared.
ufoinfo/humanoid: June, 2009

TICUL CHUPACABRAS   In April, 2008, four sheep were killed in this region of Mexico. A hairy creature was seen outside by a female witness though not, insofar as I can determine, near the sheep. It was drinking from a bucket and had a long tongue. The animal disappeared, perhaps by flying.   Locals identified the creature with the chupacabras.   Some families reported a creature resembling a bear in the area.
www.iraap.org: rosales

TIMBO   A bipedal creature with a canine appearance, sporting red hair and strong arms and claws, with a tendency to rob graves in Honduran folklore.   It was reported around the inception of the 20th Century.

TINTAGEL SEA MONSTER   A large creature, of which no further description seems available, seen off the Cornish coast in 1905.
www.paranormaldatabase.com

TIPTON MONSTER   A yellowish animal, about the size of an Airedale, attacked a boy named Frank Tomlinson in woods near Tipton. The boy was saved by his dog, which attacked the beast.   Several people claimed to have seen the animal, which was somewhat, but not entirely, like a wolf. There may have been more than one animal.
V: 251-3

TLIISH NAAT'AGD   Hopi name for a flying snake.

TLIISH NAAT'AI   Navaho name for a flying snake.

TNT BLOB In the TNT area of West Virginia, famous for its Mothman sightings, is a rather spooky place of where it is said anything can happen. Three people were once chased from the area by a strange apparition. It pursued them until they entered a car. Looking back they saw, apparently rather unclearly, a humanoid figure, both big and wide. It was coloured white.
WT: 59-61

TNT GLIDING CREATURE Another creature seen in this strange area of West Virginia is reported here. In 1973 the occupants of a car noticed a creature gliding alongside them. Its hair was shaggy, it had a huge head, it didn't appear to have wings and it maintained the speed of the car, about 65 m.p.h.
WT: 55

TOAD-FISH A monster caught near Woolwich in 1642. It resembled a toad, but had hands like fingers. It was nearly 5' in length.
www.cfz.org.uk: 31st December, 2010

TOANGINA A sort of hominid reported from the region of the Warkato River in New Zealand.
nzcryptozoology.ucoz.com

TODMORDEN CAT Big brown cat whose hind legs seemed larger than its forelegs seen on Bacup Road, Todmorden, Yorkshire, in 2011.
paranormaldatabase.com

TOOTH WORM It was formerly believed that toothaches were caused by worms. This was, however, questioned in Fauchard's *La Chirurgien Dentiste* (1728). It is now known that such worms are a myth, but there is still widespread belief in them.

TORBAY SEA MONSTER A creature reported off the coast of Devon in 2010. It looked like a sea-monster with a long neck and the inevitable comparison with a plesiosaur was made.

It was seen pursuing a shoal of fish which seem to have thrown themselves onto the beach in their fear. A photograph was taken by Gill Pearce on 27th July. There was also a photograph by Hannah Finch.

G. Oakley at first mistook the creature for a turtle, then realised his mistake. A turtle identity seems at this stage to have been completely discounted.
www.xnewsarchive.com: August, 2010; www.thisissouthdevon.co.uk

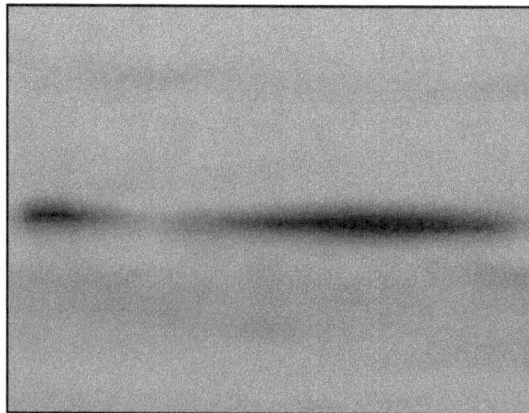

TREETON DYKE STICKLEBACK In Yorkshire folklore, this body of water held a unique species of stickleback. The story seems to be a myth, which originated with the fact that the water was at one stage so polluted only a (known) stickleback could survive in it. Happily, it is polluted no longer.
A&M 33: 31-32

TROUTMAN FLYING SERPENTS Reported from Troutman (North Carolina) in 1904. Mrs John Lipman and her family saw thirty or more snakes flying through the air.
HR: 238

TRUE GIANT L. Coleman and M. Hall feel there is a giant humanoid considerably larger than Bigfoot to which they give this name. They feel it may represent a survival of Gigantopithecus, which may be a humanoid, rather than an ape as its name implies. These giants may grow to as tall as 20'. They leave four-toed footprints, with sometimes evidence of a further toe. They are sometimes described wearing items of clothing.

While these scholars believe they are the origins of the giants of myth and folklore, the present writer feels that in certain cases, such as the jotuns of Norse myth, this is unlikely.
TG: passim

TRUSTHORPE CREATURE A large creature resembling a snake was reported off the Lincolnshire coast in the 1930s.
www.paranormaldatabase.com

TSE NINAHALEEH A monster in Navaho mythology. A woman impregnated herself with an eagle's feather and gave birth to a headless child with feathers on its shoulders. He grew into Tse Ninahaleeh, a monstrous eagle.
Wikipedia

TSUL 'KALU A sloping-eyed hairy monster of Cherokee legend. It is not clear whether it was originally a single entity, but it is now sometimes considered a species. It has been noted since 1823.
Wikipedia

TSUPA Gigantic humanoids in the lore of the Seefa Indians of British Columbia.
TG: 93

TULPAR Tulpar, who seems to be an individual, is found in the myths of Central Asia. He is sometimes thought of as a winged horse, sometimes as merely a very quick one.
Wikipedia

TUNBRIDGE WELLS APE An out of place ape seen in Tunbridge Wells, Kent, during the Second World War. A similar ape was reported recently in the same general area.
CFZ website: October 12th, 2012.

TURKISH HUMANOID This was seen to escape from a crashed UFO by a Turkish married couple in 1964. The husband attacked the creature and was knocked unconscious. The wife was unharmed. One has to ask if the Turkish authorities investigated the crash site. Perhaps they were incredulous about the whole episode.
R: 90

TURTLE MONSTER In July, 2010, a creature that looked like a turtle, but had a thin neck 2' in length and a small head was seen off the coast at Saltern Cove (Cornwall).
www.paranormaldatabase.com

TUTTLE BOTTOMS MONSTER A strange creature which has been believed to exist or have existed near Harrisburg (Illinois) in Saline County. It is a hairy creature, sometimes proceeding on four feet, sometimes on two.
www.cryptozoology.com

TUYEN QUANG PIGS Two kinds of unidentified pig have been reported from the province of Tuyen Quang of Vietnam. One has a long snout, long legs and white-striped cheeks. The other is short-faced and black.
scienceblogs.com/tetrapodzoology: December 1st, 2010

TYGOMELIA This beast was allegedly captured in Canada, north of Lake Athabasca, by a British officer named Packenham. It was described as about 5' tall but able to stretch its neck about a further two feet, leading to its being dubbed "the Canadian giraffe". It was dark brown with black spots, projecting eyes and a suggestion of horns. It has been suggested that the creature was a young moose that was mottled, as occasionally occurs, or a moose infested with ticks.
Ottawa Times: November 22nd, 1870; *P 55*

TYLOSAUR A water dinosaur that flourished 70-80 million years ago. B. Ricketts has suggested that tylosaurs just might be what lie behind Canadian reports of lake monsters.
www.mysteriesofcanada.com

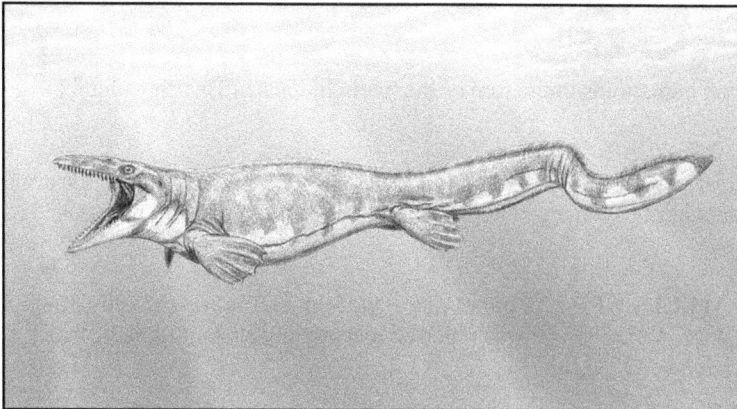

U

UKRAINIAN CHUPACABRAS   Yet another animal to which the term *chupacabras* has been arbitrarily applied.   A beast had been attacking and exsanguinations animals at villages in Vinnita. In May, 2010.

When Maxim Godonyuk heard a dog being attacked, he went to investigate and the attacking creature jumped on his back, but then retreated, perhaps scared off by approaching lights. Another witness in the area claimed to have seen a bipedal creature with striped fur and a huge head.
JHS: 1:1

UKRAINIAN CREATURE   A creature that resembled a kangaroo standing on its hind legs. It was coloured red.
ufoinfo/hominid: 2009

UMANG   Alternative name for the orang pendek.

UNDERGROUND APEMAN   This was seen in the London Underground (metro/subway) by a man named Campbell.   The lower portion of the apeman seemed to merge with the platform as though it were in some other dimension or a somewhat spectral being.   This incident supposedly took place in the 1960s.
PL: 92

UNIDENTIFIED MONSTER   This is the only name I can give to this bizarre creature, which features in a newspaper report of 1874 and one suspects a hoax is involved.   It was reported in the United States.   Its body was 6' long, its arms each 15' long, its tail 10' long with two prongs at the end, it had protuberant eyes, a horn and its four legs ended in webbed feet.
www.cryptomundo.com: January 16th, 2010

UNKNOWN NIGHTJAR   Nightjars which could not be identified have been reported from the West Indies since the 1980s.

UNKNOWN PECCARY   Unknown peccaries, perhaps belonging to different species, have been described from Ecuador and Brazil.

UNKNOWN PRIMATE   In 1995 A. Morant claimed to have discovered a painting of a possible unknown hominid amongst the cave paintings at Chauvet in France.   Others, however, have said it is a rhinoceros.
cryptozoologos.blogspot.co.uk: April 25th, 2012

UPPER LAKE MONSTER Long creature, perhaps a gigantic eel, reported from the Upper Lake in Killarney, Ireland. The wake of the creature was filmed. One of the witnesses was veteran cryptozoologist Jonathan Downes.
MAI: 55

URAYULI Hairy primates said to live in Alaska, reputed to attain a height of 10'.
Wikipedia

URDOKOTTUR A somewhat feline-looking animal in Icelandic lore. It is said to rob graves and is therefore also known as the ghoul cat.
AZ: 291

V

VACHERIE GIANT A humanoid reported from Vacherie, Louisiana, in 1891. It was almost naked, about 9' tall and covered with red hair. It ignored the witness, who plied it with questions.
JHS: 4:1

VADAKILLA MONSTER A mystery creature reported from the village of Waphere Wadi in Maharashtra, India. Traces in the form of green hair and pugmarks have been discovered. No further description is to hand.
Wikipedia

VEGETABLE MAN   West Virginia seems to be a focus for strange phenomena, but few are so unusual as the Vegetable Man. The creature is a sort of humanoid plant stalk with three green fingers having suckers and tips resembling needles. This creature was encountered by one Jennings Frederick in 1968.

The creature used its fingers to draw blood from Frederick, hypnotising him for anaesthetic purposes. It had first spoken to him in a jabbering way.
DSC: 68-69

VENTURA MARSH MONSTER   A monster referred to as the Mugwump was said to be found in this swampy area of Iowa.
globegazette.com: October 30th, 2005

VENUSIAN ANIMALS   The conditions on the planet Venus would seem to preclude the existence of any life forms there, but L. Ksanfornalitis (Russian Academy of Sciences) claims that the planetary probe of 1982 has shown three possible living creatures, resembling a scorpion, a flap and a disc.
Daily Caller: January 23rd, 2012

VERMONT GOATMAN   A creature half man, half goat, reported in Jericho in the 1960s. It is said to have subsequently gone to Mount Mansfield.
VMG: prefatory pages

VICTOR VALLEY CREATURE   This valley is in California. In October, 1996, near Victorville, a motorist returning home in the evening twilight saw a bipedal creature, whose height she estimated at 4'-4.5', standing in the road. She drove off. It pursued her on two legs. The matter was reported to the Sheriff's office. The creature had glowing red eyes, was grey and had dark spots on its chest. It also had tufts of hair on parts of its body.

In Pinion Hills, Victor Valley, witnesses saw a creature with glowing red eyes looking through the window at them in March, 1997.
FZ: 21st January, 2012

VICTORIA CREATURE   The Victoria in question is the one in Argentina. The creature was surprised by a woman at night in 2008. It was about a metre tall and extremely black. Its hands were clawed. It seemed to be interested in her chickens. She turned around to call her son and, on turning back, found it was there no more.
www.iraap.org: rosales

VILPONI   A lizard-like reptile said to live in Chiloe (Chile).
Spanish Wikipedia

VIMINI GATA   In the traditions of the Samoans, a large serpent able to crow like a cock. It was first mentioned in print in 1874.
BR: 10

VIRGINIA FLYING SERPENT   This 5' creature was allegedly shot by J.S. Dickinson in 1905.
HR: 250

W

WA'AB   Huge humanoids said to live in Sudan near the Red Sea coast.
TG: 64

WABASH VALLEY CREATURE   In the Wabash Valley in Indiana, Henry McDaniel and family heard strange scratching at their door one night. McDaniel opened the door to face a three-legged humanoid creature with a disproportionately large head and a grey body. It was 4'-5' in height and had pink eyes. McDaniel, aghast, shot it, but did it no perceptible harm. It made large bounds away. He saw it again a few nights later, standing on a railway line. All this happened in 1972. McDaniel had earlier seen strange tracks, perhaps made by the creature.
DSC: 63-64

WAHHOO   On September 17th, 1879, the *Reno Evening Gazette* reported this creature whose eyeballs were so bright that none could gaze on them. A hunting party set out after this intimidating quarry, equipping itself with huge quantities of whiskey, which I am glad to see from its spelling was proper Irish whiskey and not that Scotch stuff which has no *e.* I think we may be sure that few viewed this story seriously.

WALKENWOODS CREATURE   A number of poachers opened fire in this wood in Cheshire when they felt they were being followed. They then saw a dark shape about 7' tall. No features were discernable. In a strange way it seemed to be absorbing the darkness. This incident was reported in 2009.
P 40: 13

WALKER LAKE MONSTER   This Nevada lake is, according to Paiute Indian legend, the dwelling of two monsters, one male, the other female.

They are called *Tawaga.* A monster was described by white settlers in 1868. It is speculated that there is one monster living in the lake which lives in caves by night. The monster has been given the nickname Cecil.
FZ: January 1st, 2012

WALNUT GROVE ANIMAL   An unidentified animal which made lots of noise was in this region of Missouri in 1950. According to witness C. Shull, who had two sightings, it had the build of a bobcat, but the tail, legs and shape of the head did not conform to such an identification.
V: 337-9

WALTRON Lowland Scots name for the *each uisge.*

WANAQUE FLYING CREATURE In 1966 a boy in this area of New Jersey claimed to have seen a flying creature which seemed to have fur rather than feathers over a pond at this New Jersey location. The witness did not say it was a bird, but rather that it was "birdlike". *MNJ:* 64-65

WANI A possible Japanese alligator. It may be the same species as the Chinese alligator (below). *ACC*

WARUSKA Giants in the folklore of the Iowa Indians. *AE:* 168

WASHINGTON PTERODACTYL There have been various reports of pterodactyls in the rainforest in Olympic National Park (Washington).

WATER BABOON A South African cryptid, reported from the Free State. It is said to resemble Australopithecus, which zoologists believe to be an early ancestor of man. *About.com: paranormal phenomena:* August, 2005

WATER DOG   An animal reported from Manitoba. It has big ears, short legs and a short snout. It is coloured brown.

WATER GOBLIN   These creatures are said to be hairy diminutive humanoids with round heads. They are rumoured to be found in Lake Vedbzero, Russia. Nobody reported them before the 1990s.
AZ: 324

WATER TIGER   Man-eating animal living in rivers in the beliefs of the Sumu Indians of Nicaragua. We should imagine *tiger* here means *jaguar*, referred to as *el tigre* in Latin America.
CR: 2:1

WAUKESHA ANIMAL   A strange-looking animal photographed in Waukesha County (Wisconsin). It has not been certainly identified and some have suggested it is a chupacabras. A red fox with mange has also been suggested.
www.wisn.com: October 22nd, 2010

WAWANAR   This may be identical with the ropen, mentioned in the author's *Dictionary of Cryptozoology* (2004). It is known by this name on Pilio Island, Papua New Guinea, and, if not identical with the ropen, may yet be a surviving kind of pterosaur.

WENTSHUKUMISHITEU   A water monster of Eskimo legend. It protects other creatures from human hunters, whom it will devour. It lives under the ice floes. It can also force its way along underground.
Wikipedia

WEREDOG   There was an historical character in America and legend has it that he sometimes turned into a dog. The person concerned was known as Railroad Bill. Some said he was a Robin Hood style figure. His identity is uncertain. He may have been called Morris Slater or Bill McCoy.
DWF: 360; *Encyclopedia of Alabama* (Internet)

WEST BOLDON GIANT BIRD   A large unidentified bird seen by M. J. Hallowell at this location in the north of England, where he lived at the time. It was 3' high, milky-brown in colour and had the legs of a wading-bird. It concealed itself in a bush and, fearing it was going to attack his son and another child, Hallowell yelled a warning. It emerged from the bushes and the children fled. Staff at the local bird sanctuary said it did not conform to the description of any of their resident avians.
N&T: 27-28

WEST COAST SPOTTED KIWI   A New Zealand bird, thought to have become extinct about 1900, but it may never have existed. Reports may have been based on a hybrid.
Wikipedia

WEST ORANGE CREATURE   In 1924 a Mrs Vincent had a curious encounter with a creature at this locale in New Jersey. A creature with a deer's head that jumped about like a rabbit and had fiery eyes approached, but does not seem to have wrought any harm. A policeman also saw the creature.
MNJ: 70

WEST VIRGINIA LION   An out of place lion reported in 2007.
MAWV: 97

WEST WEMYSS SEA SERPENT   This creature, which had a long neck with an equine mane and large eyes, was seen off the Scottish coast in 1939. Its length was estimated at 6m.
www.paranormaldatabase.com

WHALE-FISH   Creature said to be found in Lake Myllesjon, Sweden. Reported from the middle of the 19th Century, it seemed to be seen quite often in the first quarter of the 20th. A motorway/freeway beside the lake at present may discourage the creature from surfacing.
P 56: 34

WHATZIT   A creature reported from Buck Mountain (Vermont) in the 1950s/1960s. It looked like a bipedal goat without horns. Its hair was coloured whitish-grey.
VMG: 49

WHIRLEPOOL   Three fish so called were captured on 7th October, 1552, at Gravesend (Kent). The size of these fishes was considerable.
www.kentmonsters.blogspot.com

WHITE BAT-CREATURE   These were creatures about 3' long that emerged from a hole in Florida. A vespertilian membrane stretched from arms to legs. They were observed in 2001.
CFZY 2009: 170

WHITE CAT   This animal has been reported from Isiolo, Kenya, killing dogs and sheep. Its species is unidentified. It appears to belong to a group of cats, but whether they too are white is unclear.
www.cryptomundo.com: July 2nd, 2012.

WHITE COUNTY SERPENT   Several persons, including one H.C. Cotton, allegedly saw this serpent in Tennessee in 1899.
HR: 239

WHITE DEVIL   A white creature about 5'8" tall with sharp claws which attacked a hunter in West Virginia.
WT: 73

WHITE HAIRY THING   The woman who was the source of this story told how her mother had seen a white hairy thing when looking out of her window at night at her West Virginia

home. It was proceeding on all fours. She assumed the incident occurred in the 1970s.
WT: 74

WHITE CREATURE This was seen at Napadovka, Ukraine, in December, 2009. It resembled a kangaroo. There had been animal mutilations in the area. *See also* **Red Beast.**
HSR: 2009

WHITE LION A white-coloured lion that seemed to have phantasmal characteristics, seen by the father of author Kurt McCoy in West Virginia. While white lions have been bred in the past, it is not certain that this is the same sort of creature.
WT: x

WHITE-FACED MACAW Unknown species of macaw painted by Roelent Savery (16th Century). It had a black half-collar, green wings and a yellow tail (right). The centrepiece of the painting is a dodo, so the island where it is depicted is probably Mauritius.
Shukernature: March 24th, 2011

WHITEHEAD'S SWIFTLET (*Aerodramus whiteheadi*) A bird of the Philippines, whose continued existence is doubtful, but not impossible. Apparently very similar to the Philippines swiftlet (pictured here)

WHORTLECHORT The name given to a hairy humanoid reported from Kentucky in 1965.
It was estimated at 7'- 8' in height. It may have been a Bigfoot. The witnesses, not having heard of Bigfoot, coined this name for it.
I: 243

WHOWIE There seems to have been some doubt amongst Australian Aborigines regarding what the whowie is - a six-legged lizard or a six-legged insect with a frog's head. It apparently lived in a cave by the Murray River and humans featured in its diet. One legend says the whowie was imprisoned in its den with brushwood and then smoked to death.
DSC: 117; *www.cfz.org.uk*

WIECBORK CREATURE This was spotted in the Kujawsko-Pomorskie area of Poland on May 28th, 2010, by a motor-cyclist. He said it seemed to have a watery film covering it. No further details are available.
JHS: 1:4

WILD LITTLE MEN These are believed in by the Wailaki Indians of California.
AE: 120

WILDERNESS WOLF A creature in the beliefs of the Indians of British Columbia. G. McIsaac opines that it represents a surviving population of the supposedly extinct Dire Wolf (*Canis dirus*), pictured left from a mural at the La Brea tar pits, a relation of the grey wolf which is supposed to have become extinct 10,000 years ago.
BH: 62

WILDMAN OF ENON Local name for Bigfoot type entity in the region of Enon (Ohio).

WILLOUGHBY GIANT FISH Lake Willoughby (Vermont) is said to be home to an unidentified giant species of fish, large enough to swallow a man. The creatures have been reported by divers.

VMG: 83

WILLOUGHBY WISP A creature supposed to dwell in Lake Willoughby (Vermont). It has appeared in the lake in the form of humps and there is even a record of one being killed in the 19th Century. The lake was home to more than a single animal, however, as one was seen from Westmore in 1986. There have been reports of huge eels in the lake and these may be identical with the Willoughby Wisp.
VMG: 76-78

WINFIELD ROAD CREATURE A white animal, much larger than a dog, was seen on this road near St Albans (West Virginia) by motorists. When first seen, it was on all fours, then it rose onto its hind legs and ran into the woods.
WT: 24

WINGED CREATURE This looked like a deer and had cloven hooves and wings when it was encountered by a motorist in Lake County (Oregon) in 1996. It flew away.
SM: 83

WINGED MAN This large feathery creature was seen by at least one witness, apparently in Kitsap County (Washington).
cryptids and folk beasts in Washington?: June 1st, 2006

WINONA BEAST In 1919 a strange beast was seen in the vicinity of Winona (Minnesota). It was described as being the size of a yearling calf, grey and striped.
V: 330-1

WOMAN-FACED MACAW This singular member of the avifauna survived the Flood and interbred with its two male human survivors, according to the Canaris Indians of Ecuador.

WONDERFUL BIRD A bird that could eject and retrieve its eyes in the mythology of the Blackfoot Indians.
BLT: 153

WOODBURY WATER WITCH This is a strange amphibian said to be found in Woodbury Lake, formerly Sabine Lake (Vermont). It has "recessed" eyes, which I take to mean eyes somewhat sunken into its body, a scaly body and a tail reminiscent of a web.
VMG: 75

WOODMAN A creature seen in 1985 by Arlene Tarantino in the Green Mountain National Forest (Vermont). It looked human, but moved on all fours. Its body appeared youthful, but its face looked like that of an old, wrinkled man.
VMG: 42

WOOG An animal described as having a long bushy tail, large feet and small ears with a tendency to hunt livestock by night, which was active in Georgia in 1895.
V: 206

WOOLLY CANID   Four canine animals, too woolly to be wolves, attacked dogs belonging to T. Ledbetter in Faulkner County (Arkansas) in 2008.   They backed Ledbetter himself against a fence, but he was rescued by a gallant basset.
V: 154

WOULDHAM HUMANOID   A humanoid reported from Kent, first from the area of Wouldham, in the early 1900s.
P 55: 72-3

WUCHARIA   A canid, probably a jackal or wolf, of unknown species reported from Eritrea. It may be a form of the Arabian wolf (*Canis lupus arabs*) which is pictured below.
Wikipedia

X

X Jacqueline Roumeguere-Eberhardt (1927-2006) was an extraordinary anthropologist who was the wife of a Masai warrior. She heard various reports of different kinds of humanoids in Kenya to which she gave the designation - not the name - X. She described five types of X from 33 encounters in 11 forests:-

- **X1** Large and blond, carrying a spiked club. This creature was seen by 31 adults and two children in eight forests.
- **X2** This one has no hair and is coloured white. He is nocturnal and emits screams. There were a dozen witnesses.
- **X3** Tall, black, with white hair that reaches foot level. He wields a club, but, when he kills a buffalo, he drinks its blood, leaving the remainder. There were five witnesses.
- **X4** He has a large head and has been reported with a female and young.
- **X5** He was like X4 but had a bow and arrows. Some Masai acquired this and a bag for nettle collecting.

The present writer is not certain whether we should distinguish between X4 and X5. X4 is supposed to eat tubers, mushrooms, berries and, according to one source, carrion. The bow would indicate X5 hunts meat, but maybe X5 is only an X4 who happened to be seen with a bow.
Bord: 83; *Des Moines Register:* October 23rd, 1978; *H:* passim

XIAMEN SEA MONSTER The city of Xiamen, China, was, in the 1930s, called Amoy. In those days a monster with a human-like head was observed in the distance by sundry locals.

Descriptions were vague.
www.cfz.org.uk: 24th November, 2011.

XIEZHI A fire-eating dog that looks like a lion in Chinese and Korean mythology.

A xiezhi can also be thought of as a goat with a single horn, with which it attacks liars.
Wikipedia

XOLCHIXE In the jungles of South America, this animal is rumoured to exist. It is said to be the size of a lion, to be carnivorous and sometimes to be found in the trees. Natives aver it is a kind of sloth.
Wikipedia

Y

YARRA ANIMAL A creature reported from Australia in 2010. It was the size of a Rottweiler and its back did not appear to slope.
P 46: 13

YEL'TSO Anthropophagous giant of Navaho lore. A sighting of three specimens was reported from New Mexico in 2005.
TG: 97-8

YEMASSEE THING An animal reported in South Carolina. It was described as weird. It was four-legged, 5' long, 4' high, with a bushy tail. Its colour was a yellow/brown.
V: 594

YI-NIAO Huge bird found in the lore of China.
CMS: 254

YOLI A large snake with two legs, reported by the natives of Cameroon. It is said to generate electricity and to stretch itself upward higher than the tallest tree.

YU According to the Chinese *Shan Hai Jing*, a creature resembling an ape. Its eyes are scarlet, its tail is long. It was found in the mountains bordering the Yangtse river.
CMS: 199

Z

ZARIA CREATURE A mystery creature in the Ukraine which killed three oldsters and many rabbits. It does not appear to have been observed.
JHS: 1:1

ZHEN A Chinese mythical bird with green tips on its feathers. It is poisonous, because it eats the heads of snakes. Its abdomen is purple, its beak bright red. Its poison is termed *zhendu*.
Wikipedia

ZIMBABWIAN MERMAID A white woman with a scaled fish-tail was reported by M. Batau in January, 2000, sunbathing by the Hunyani River. He saw her in all three times, on one occasion feeding a baby.
www.iraap.org: Rosales

ZIONSVILLE CREATURE A flying creature of unidentified species seen at this Indiana location by two boys. It had an angular face and was coloured scarlet. It was about the size of a pigeon. J.D. Whitcomb opines it might have been a pterosaur.
LPA: 39

ZORRO-VIBORA Spanish name for the nurufilu, which see. The name means 'fox-viper'.

Printed Sources

A&M   *Animals and Men*

ABC   Williams, M./Lang, R. *Australian Big Cats* Hazelbrook, 2010

AE   Roth, J.E. *American Elves* Jefferson, 1997

AZ   Shuker, K.P.N. *Alien Zoo* Woolfardisworthy, 2010

BH   McIsaac, G. *Bird From Hell* n.p., 2010

BLT   Grinnell, G.B. *Blackfoot Lodge Tales* London, 1893

Bord   Bord, J. and C. *The Evidence for Bigfoot* Wellingborough, 1984

CFZY   *CFZ Yearbook* (year numbers supplied in text)

CMS   Birrell, A. (tr.) *Classic of Mountains and Seas* London, 1999

DFD   Glavin, T. *Death in Dimlahamid* Vancouver, 1990

DSC   Shuker, K. *Dr Shuker's Casebook* Woolsery, 2009

DWF   Jones, A. *Larousse Dictionary of World Folklore* Edinburgh, 1995

EB   Heinselman, C. (ed.) *Elementum Bestia* Peterborough (NH), 2007

FS   *Flying Snake* (Richard Muirhead)

FT   *Fortean Times*

FUW   Newton, M. *Florida's Unknown Wildlife* Gainesville, 2007

GNB   Forbes, R.A. *Gaelic Names of Beasts* Edinburgh, 1905

GW   Williams, P. *Mystery Animals of the British Isles: Gloucestershire and Worcestershire* Woolsery, n.d.

H   Roumeguere-Eberhardt, J. *Hominides non-identifiés de forets d'Afrique* Paris, 1990

HG   Tomkins, J.L. *Haunted Greece* Athens, 2004

HLOH   Westerveldt, W.D. *Legends of Old Honolulu* Boston, 1915

HM   Beckwith, M. *Hawaiian Mythology* New Haven, 1940

HR   Clark, J. *Hidden Realms, Lost Civilizations and Beings from Other Worlds* Detroit, 2010

I   Nunnelly, B. *The Inhumanoids* Woolfarthisworthy, 2011

K   Arnold, N. *Mystery Animals of the British Isles: Kent* Woolsery, 2009

Leland   Leland, C. *The Algonquian Legends of New England* Boston, 1884.

LM   Cardenas Alvarez, C. *El Libro de Mitologia* Puntas Arenas, 1997

LPA   Whitcomb, J.D. *Live Pterosaurs in America* n.p., 2009

MAI   Cunningham, G./Coghlan, R. *Mystery Animals of Ireland* Woolsery, 2010

MAL   Arnold, N. *Mystery Animals of the British Isles: London* Woolsery, 2011

MAP   Gable, A. *Mystery Animals of Pennsylvania* Woolsery, 2012.

MAR   Gray, L.H. (ed.) *Mythology of All Races*

MAWV   Guiley, R.E. *Mystery Animals of West Virginia* Mechanicsburg, 2012

MLBOB   Skinner, C.M. *Myths and Legends Beyond Our Borders* Philadelphia, 1899

MNJ   Coleman, L./Hallenbeck, G. *Monsters of New Jersey* Mechanicsburg, 2010

Mooney Mooney, J. *Myths of the Cherokee* Washington, 1900

MS   Holder, G. *The Guide to Mysterious Stirlingshire* Stroud, 2008

MSB   Codd, D. *Mysterious Somerset and Bristol* Derby, 2011

MT   Gerhart, K./Redfern, N. *Monsters of Texas* Woolsery, 2010

MTBM   Spence, L. *Minor Traditions of British Mythology* London, 1948

MW   Godfrey, L. *Monsters of Wisconsin* Mechanicsburg, 2011

NI   Vaudrey, G. *Mystery Animals of the British Isles: Northern Isles* Woolsery, 2011

NM   Thorpe, B. *Northern Mythology* Ware, 2001

N&T   Hallowell, M.T. *Mystery Animals of the British Isles: Northumberland and Tyneside* Woolsery, 2008

P   *Paranormal* (magazine); ed. R. Holland

PL   Arnold, N. *Paranormal London* Stroud, 2010

PaM   Knappert, J. *Pacific Mythology* London, 1992

R   Steiger, B. and S. *Real Aliens, Space Beings and Creatures from Other Worlds* Canton (Mississippi), 2011

RMP   Wilson, P.A. *Really Mysterious Pennsylvania* Mechanicsburg, 2010

SI   Gordon, S. *Silent Invasion* n.p., 2010

SIM   Boryan, M. *Solomon Islands Mysteries* n.p., 2009

SKM   Newton, M. *Strange Kentucky Mysteries* Atglen, 2010

SM   Newton, M. *Strange Monsters of the Pacific Northwest* Atglen, 2010

TSW   Redfern, N. *There's Something in the Woods* San Antonio, 2008

V   Arment, C. *Varmint* Landisville, 2010

VMG   Citro, J.A. *Vermont Monster Guide* Hanover, 2009

WI   Vaudrey, G. *Mystery Animals of the British Isles: the Western Isles* Woolsery, 2009

Wilson   Wilson, H.E. *The Lore and Lure of the Yosemite* San Francisco,
        1922
WT   McCoy, K. *White Things* Morgantown, 2008
WW   Thomas, L. *Weird Waters* Woolsery, 2011

Internet Abbreviated Sources

ACC   D. Drinnon Additional Cryptozoological Checklist on CFZ
         Website (www.cfz.org.uk)
BR   *Biofortean Review*
CSY   *Cryptomonsters Sneaking Up on You*
FZ   *Frontiers of Zoology* (D. Drinnon)
HMR   *Humanoid Sightings Reports*
JHS   *Journal of Humanoid Studies*
PM   *Patagonian Monsters* (A. Whittall)
VIC   *Virtual Institute of Cryptozoology*

Snow Leopard: Spirit of the Mountains

Raheel Mughal

Introduction

The snow leopard has to be my favourite animal. That being said, I wonder why this beautiful cat is never given the attention that it deserves by natural historians, conservationists and the like. Instead, greater focus is given to the giant panda, the great whales, polar bear, tiger, mountain gorilla, orang utan, the lemurs of Madagascar, sharks, and the various multi-coloured frogs of Central and South America. All animals are amazing, don't get me wrong, but conservation is about the big picture (the entire ecosystem in which a target animal lives). Conservation to me (and I'm sure my dear readers will agree) involves steps to safeguard the entire local environment in which a target species lives because everything in a biome is linked, from lichen to apex predators, so naturally focusing on one animal neglects the big picture, and the big picture is what we are fighting to protect.

Photo by J. W. McLelian]   [Highbury.
SNOW-LEOPARD, OR OUNCE.
This is a striking portrait of a very beautiful animal. Note the long bushy tail, thick coat, and large eyes.

Nevertheless, the snow leopard is important in more ways than one; this elusive big cat is a symbol of Central Asia, of the remoteness, and the fragility of living on the edge on top of the world. This snow cat is also an important cultural ingredient to the lives of many indigenous people inhabiting the Central Asian Plateau, or the Himalayan mountain range as it is known.

Readers may think that they know a thing or two about the Spirit of the Mountains, but after reading my research I am sure you will learn something new. The emphasis of this paper is to highlight the plight of this cryptic cat. I hope and pray that, armed with this new found knowledge, readers may be encouraged to think up some ingenious ways to help to protect this beautiful cat from the ever-present pressures that threaten to exterminate it. God forbid.

Physiology

The snow leopard, or ounce, was once a legendary cat that stalked the mountains and alpine meadows of Pakistan's Karakoram range.

However, until fairly recently zoologists and naturalists have begun to uncover more and more about this elusive snow cat. So much so that the snow leopard has been given legendary status once more as the national predatory animal of Pakistan. It is commonly known as the "Barfani Chita" by the natives.

Snow leopards are approximately, 1-1.3m (3 ¼ to 4 ¼ ft) long and weigh 25-75 kg. (55-165 lbs), respectively. This makes the snow leopard the fifth largest big cat species after the tiger, lion, jaguar and leopard.

Snow leopards, despite their name, are not closely related to leopards. In fact a large majority of biologists and scientists believe that the snow leopard represents an archaic species of big cat distantly related to lions and jaguars. Snow leopards have large dark rosettes on a ground of pale grey or creamy smoke-grey. They have the longest tail of any cat, which can sometimes exceed one metre (3ft) in length.

These cats are solitary in nature except during courting. Snow leopards produce one to five cubs during the spring and these are cared for by the female; the male parts ways after this period.

Snow leopards have large territories above the snowline. They are expert ambush hunters that prey on a wide variety of animals, which range from, but are not limited to, marmots, mountain vole, Himalayan mole, bharal, asiatic mouflon, urial or blue sheep, Himalayan tahr, ibex and - their favourite - the markhor.

Snow leopards wait perfectly still in ambush on top of a ledge or rocky outcrop waiting for a prey item to walk past, and when they catch sight of a potential target a mad pursuit is set in motion. If the snow leopard is successful then it will haul the carcass of its unfortunate victim up to its den (usually a cave or crevice situated in a rock face) for later consumption, or more usually (if it happens to be a mother) for the cubs. Like all predators not all hunts are successful, but this intelligent cat has survived for millions of years having a varied diet; in the

**Snow Leopard - *Uncia uncia* - at the Lisbon Zoo
(Alfonsopazphoto /Wikimedia Commons)**

cold conditions where it lives it doesn't pay to be a fussy eater. The main threat facing these beautiful cats is habitat loss (the encroachment of livestock and human dwellings) and this often puts the snow leopard in danger from angry villagers who cite the snow leopard as the main cause of their subsistence problems.

The other major problem facing the survival of the snow leopard is its beautiful woolly coat, which - incidentally - is the main cause of its demise. Snow leopards are hunted extensively for their luxurious coats which is an illegal practice adopted by unscrupulous fashion

PLATE VI.

designers. It is a sad irony that a coat that protects the snow leopard from the elements is also its greatest enemy.

That being said, all is not lost. For instance WWF Pakistan is working closely with other environment agencies to stop this barbaric practice once and for all.

Snow leopards are the most elusive big cat in the world. In the wild, wildlife enthusiasts have to wait weeks or even months in order to catch a glimpse of this spirit of the mountains. Their secretive nature means that their exact numbers are unknown, but according to Pakistan Natural History Museum it has been estimated that between 3,500 and 7,000 snow leopards exist in the wild.

Mythology

The Wakkan people who live near the border between Pakistan and Tajikistan (the Wakkan Corridor as the area is known), talk of the Mergichen, a race of spirit people who inhabit the plateaus of the mountains surrounding the area. They are said to possess the head of a snow leopard, as well as the hands, feet and tail of one, but have the body of an adult human wearing the same kind of dress as that of the Wakkan people (reminds me of King from the Tekken video games series). They can morph into an adult snow leopard if they so wish. They are said to be a lucky omen for anyone who gets to see one, and they are said to come down from the mountains during the winter when food is scarce to feed on mountain goats, such as ibex and markhor (this, the Wakkan say, is a good sign that the Mergichen are pleased with

Snow Leopard *Uncia uncia* at the Louisville Zoo (Ltshears/Wikimedia Commons)

how humans are treating the land). When pleased, the Mergichen are said to use their magical powers (other than those discussed above), such as granting wishes to help with a good harvest in specific areas if they so wish. Nevertheless, if they are angered then the full fury of their rage will be set in motion, leaving a trail of destruction on anyone unfortunate to have this ill omen dropped onto them.

The Mergichen could well be the Wakkan equivalent of the European Werewolf mythos. Nearly every culture on earth seems to have a being that incorporates human characteristics with a local "spirit" animal, examples include: Werebears, Werehyeanas (Africa), Weretigers (India), and Land Otter Men (N.America), which I talked about in detail in the *CFZ Yearbook 2012* chapter Mystery Animals of Native American and Innuit Mythology, respectively. There seems to be some sort of human desire to incorporate animal traits from locally known predatory animals and grafting them onto the human form. This could have started early on in prehistory when there were regular skirmishes with warlike invading tribes, and the animal-human hybrid design was probably designed as a fear inducing tool so that invading tribes would lose interest, back off and retreat.

References

* Lonely Planet Guide (2004), *Pakistan & the Karakoram Highway*, 6th ed, Lonely Planet Publications
* UK.Shaw. I, (1998), *Pakistan Handbook*, 2nd ed, Moon Publications, UK
* Pakistan Natural History Museum (visited April 2011)
* www.snowleopard.org
* www.bbc.co.uk/wildlife
* www.wildlifeofpakistan.com
* Himalayas – BBC Natural World (aired November 2012)
* Mountains – Planet Earth (aired December 2012)

The man-eating plant of ye olde England

Glen Vaudrey

It is perhaps surprising to learn that the there could be a man-eating plant and that there may be one alive and lying in wait in the British countryside. I hope with the article to show you that sometimes the strangest things are not that far from where you live. Even as write this I am no more than 12 feet from one of these plants. Before I reveal the culprit in all its leafy glory I will take you on a small tour of reported man-eating plants.

Perhaps the best known of man-eating plants are those from literature, film and television; one plant that has managed to appear in all three forms is the triffid. This dangerous plant was active, that is it's a walking plant. The triffid is of course made up despite those traffic signs one occasionally sees stating 'heavy plant crossing' which actually refer to pieces of large earth moving equipment on their travels. The triffid was a creation of John Wyndham and it is never made clear where these walking predatory plants came from, some suggest that they were the results of tinkering with nature by sneaky Soviet botanists or perhaps that they had come from space; the triffid might not have done but the next to plant certainly did. In 1965 we find *The Avengers* in the form of Steed and Mrs Peel taking on the Man-Eater of Surrey Green, an alien plant that had landed in England and then took a number of horticulturists hostage. What strange ideas these alien plants have.

There was more horror to come in the film *The Little Shop of Horrors*, here we have a misguided fellow crossing a butterwort and a Venus flytrap leading to a number of folk being fed to the plant, disturbingly named Audrey junior; if that wasn't scary enough in the 1980s remake the plant not only ate people it also sang.

While the above fictitious plants may all be easily discounted from the search there are other plants said to be out there looking for a person for lunch.

One of the most mentioned is the Ya-te-veo and while it might sound like the latest signing for Manchester City it is said to actually translate as 'I see you'. This mystery man-eater was said to have been found, and for all we know may still be found, in the jungles of Central and South America. The first mention of it appears in *Sea and Land* by J.W. Buel in 1887.

'travellers have told us of a plant which they assert grows in central Africa and also in south America, that is not contented with the myriad of large insects which it catches and consumes, but its voracity extends to making even humans its prey. This marvellous vegetable minotaur is represented as having a short, thick trunk, from the top of which radiate giant spines narrow and flexible, but of extraordinary tenaciousness, the edges of which are armed with barbs or dagger like teeth, instead of growing upright or at an inclined angle from the ground and so gracefully are they distributed that the trunk resembles an easy couch with green drapery around it. The unfortunate traveller, ignorant of the monstrous creation which lies in his way, and curious to examine the strange plant, or to rest himself upon its inviting stalk, approaches without a suspicion of certain doom. The moment his feet are set within the circle of the horrid spines, they rise up like gigantic serpents, and entwine themselves about him until, he is drawn upon the stump, when they speedily drive their daggers into his body and thus complete the massacre. The body is crushed and every drop of blood is squeezed out of it and becomes absorbed by the gore-loving plant, when the dry carcass is thrown out and the horrid trap set again.'

Quite a plant as the accompanying picture will show you when it's in action, and a comfy seat when it isn't. The idea of evolution coming up with such a design would be impressive, perhaps its one for the intelligent design folks but personally I would suspect the hand of IKEA behind it.

The next man-eating plant to appear on the scene was to be found in Madagascar. It was in 1878 that a German by the name of Karl Liche told a tale of a monstrous plant, pygmies and sacrifice. Now that does sound like it has all the ingredients for an Amicus production from the late 1960s but as far as I know it hasn't been filmed yet.

Liche would recall that by the bend of a small stream in the depths of a forest there was the most peculiar tree to be found. He describes it as follows.

'I have called it 'Crinoida' because when its leaves are in action it bears a striking resemblance to that well–known fossil the crinoid lily-stone or St. Cuthbert's head. It was now at rest, however and I will try to describe it to you. If you can imagine a pineapple eight feet high and thick in proportion resting upon its base and denuded of leaves, you will have a good idea of the trunk of the tree, a dark dingy brown, and apparently as hard as iron. From the apex of this truncated cone eight leaves hung sheer to the ground. These leaves were about 11 or 12 ft long, tapering to a sharp point that looked like a cow's horn, and with a concave face thickly set with strong thorny hooks. The apex of the cone was a round white concave figure like a smaller plate set within a larger one. This was not a flower but a receptacle, and there exuded into it a clear treacly liquid, honey sweet, and possessed of violent intoxicating and soporific properties. From underneath the rim of the undermost plate a series of long hairy green tendrils stretched out in every direction. These were 7 or 8 ft long. Above these, six white almost transparent palpi [tentacles] reared themselves toward the sky, twirling and twisting with a marvellous incessant motion. Thin as reeds, apparently they were yet 5 or 6 ft tall.'

Liche then recounted how he watched a gathering of local tribesmen use spears to force one of

THE YA-TE-VEO, OR CARNIVOROUS PLANT.   476

their accompanying women folk to climb the trunk of the tree. It would not have taken long for her to reach the top of this most peculiar of plants. She was then told to drink the treacle-like liquid that the monstrous plant produced. Even before you read the final part of Liche's statement you know this isn't going to end well for the lady in question.

'The atrocious cannibal tree that had been so inert and dead came to sudden savage life. The slender delicate palpi, with the fury of starved serpents, quivered a moment over her head, then fastened upon her in sudden coils round and round her neck and arms; then while her awful screams and yet more awful laughter rose wildly to be instantly strangled down again into a gurgling moan, the tendrils one after another, like green serpents, with brutal energy and infernal rapidity, rose, retracted themselves, and wrapped her about in fold after fold, ever tightening with cruel swiftness and the savage tenacity of anacondas fastening upon their prey. And now the great leaves slowly rose and stiffly erected themselves in the air, approached one another and closed about the dead and hampered victim with the silent force of a hydraulic press and the ruthless purpose of a thumb screw.

'While I could see the bases of these great levers pressing more tightly towards each other, from their interstices there trickled down the stalk of the tree great streams of the viscid honey like fluid mingled horribly with the blood and oozing viscera of the victim. At the sight of this the savage hordes around me, yelling madly, bounded forward, crowded to the tree, clasped it, and with cups, leaves, hands and tongues each obtained enough of the liquor to send him mad and frantic. Then ensued a grotesque and indescribably hideous orgy. May I never see such a sight again.

'The retracted leaves of the great tree kept their upright position during ten days, then when I came one morning they were prone again, the tendrils stretched, the palpi floating, and nothing but a white skull at the foot of the tree to remind me of the sacrifice that had taken place there.'

And that, it appears, is the last sighting of this strange plant. Many have been to Madagascar looking for it in the ensuing years but without success. There are a number of reasons to suspect the validity of this plant but one particular thing that I have found odd about it is why a man-eating tree of around ten foot tall would prey on pygmies, had its chosen food been unlucky parachutists I could see it at least evolving.

There are other man-eating plants said to be out there making the jungles of the world such dangerous places.

As you may have gathered by now despite the number of reported man-eating plants around the world no one has yet produced one of these let alone turned up at the RHS Chelsea Flower Show with one.

That is not to say there are not carnivorous plants to be found, there are in fact a number of different types.

Perhaps one of the most well known is the Venus flytrap which catches its prey with rapid moving leaves. Once the insect is trapped the leaves close tightly forming a 'stomach' in which digestion occurs.

Then there are the pitcher plants that trap their prey in a rolled leaf that contains a pool of digestive enzymes or bacteria which not only drowns the victims but also digests them.

We have nature's very own plant flypaper trap the sundew with its very sticky mucilage covered leaves. The poor insect becomes stuck in the leaf dying of exhaustion or asphyxiation as the mucilage clogs the insect up. The plant then releases enzymes that both dissolve the insect and free the contained nutrients, the nutrient soup is then absorbed through the leaf surfaces.

There are also the bladderworts with a bladder that generates an internal vacuum that suck the prey in. They occur in fresh water and wet soil as terrestrial or aquatic species. In the active traps of the aquatic species the prey brush against trigger hairs connected to the trapdoor. This bladder is under negative pressure in relation to its environment so that when the trapdoor is mechanically triggered the prey, along with the water surrounding it, is swept into the bladder. Once the bladder is full of water the door closes again, the whole process taking only ten to fifteen thousandths of a second, which you have to admit is pretty quick.

There are also corkscrew plants which hunt with a lobster pot-style trap in which the prey is forced towards the digestive organ with inward pointing hairs that don't allow it to turn around.

So yes, carnivorous plants do exist but unless you are really small you have nothing to fear as these plants are on the whole rather tiny, certainly not man-eaters. Small or not these plants are classed as carnivorous for not only do they trap their prey but they digest them, it's a good job none of them have managed to grow very large or come to think of it learned to walk. There is another group of plants that the man-eater of ye olde England could well belong to and that is the protocarnivorous plant. A protocarnivorous plant is best summed up as a plant that can trap and kill its prey but lacks the ability to either directly digest it or to absorb nutrients from it. It is to this group of plants that our mystery man-eater belongs and it is probably time that I revealed its identity, it is none other than the humble blackberry bush also known as the bramble. It seems that while Victorian explorers were busy scouring the globe for man-eating plants, a likely candidate could be found back at home not hidden in the middle of dark lonely forests creeping up on pygmies, but in actually contact with unwitting people as they picked the berries from this plant.

For those not familiar with the blackberry bush it is a tangled prickly shrub with thorny stems. Growing throughout the UK this seemingly harmless plant provides fruit that can be harvested in late summer and early autumn. But it is this tempting fruit that holds the key to this plant's dark secret. To catch its prey the blackberry bush entices you into danger with nothing more that a tasty berry as bait.

While its place as a man-eating plant has not up to now been recognised the blackberry bush

βᾶτος

والتوث
عليق

πασιγνωέειως, τυφτέχροιεῖσιστεῖσει, βᾶτουτοῖ
φῖται ἀδὴ τοῦλέα, ἐμφανοῦαὐτῆ, κειλιάπτοῖ
τσοι πτεσρᾶσει, ὁ ρομεῖσι γᾶλωακλδοω
δ᾽ πεσιεῖσει, δαλματι κἄσετη, κειλιωςλδ᾽εσι δ
δ᾽ ἀφρονιλφᾶλε, τ᾽αρμαικτοσικτοῖ δύσμαως
κειλοχ πῶτεσι ἔτσι δα̇ῦςεῖσιεσιτοῖ σῦ κιρῖδας
βᾶτσιλδ᾽ κειλοφεῖ νῶς προ πᾶσετ᾽, ὁκεσ
Δ ωμαῖ, κειλαμεσηγη κειλι ἔτ᾽πλατμοῦλῶ
τῶ φῶσε, κειλτὸσουκ ἔκοπτει, μᾶνῆ δι᾽κσι
λᾶμαι ἐπιτῆ τσι ἀρ μᾶζε ὁδ᾽ χυλός αὐτῆςεκσιβι
ρολαιτουκεμφεν τσι ἀτσοῦ φῶλαι, σνι τεφετσίεσικτοῖ
ρμεῖσι τσοῖ τσοῖ, τσῦλα κεσρ μεσλτικῶτσοῦμλσι
τπελεῖσιντσῦ, δ᾽λεᾶ σσμαλτικὸς δ μεσζει ἰσποῖ
δ κειλκοιλῖκ, σῦμσνεσι πεισερσ δᾶ ἑμιερσ
ἴσηδ᾽τσι σιδ᾽αὐτῆς, τσοκσιειῖοσι, κειλτᾶω
ἐστιτῆ·   πολλωσι

βᾶτσιτσδ᾽τσοσ, κειλικσι πτ᾽ κειλ ἔιλκσι συ
ἀπτιικει μεῖτ᾽ γτσοι τσιτῶ τσιωσι ὁ πτᾶθ᾽τσνει
φῖλαι, κειλτσοῦσλσιτσι, ἐκειλσιφᾶνει κσι τῶ
τσῦ ὁ πλσι θσοσ, μεῖτι μετσοι, ὁστι κειλὁλλασκσι
μᾶνοι, ἀφρσιεταῖσι, κειλτσιτσνεῖσι ῆτσῶ μαλτᾶη
ρστσνεσιφτσι δῇ κτ᾽ ὑσιτῆ ἀρσι σιτσι λσιθσι
κειλτσι σμεσδ᾽καρ πτ᾽σῆ ὁτσι πτεσσ, μεῖτσι
ἴστσο, σνμαιεῖδ᾽, ὁθλᾶ   ὑμσιρῖ
ἰ μελσιτικι, ἐτι
ξμσιελτικῶτσεσεσσ
πῦτσεσφατσι, τσιϊσι
τσῖσε κειλὁρσατσι
τσσειγσ, δᾶλσιῖ, κειλσσιῖτῖ
αἰμακτσιστικτσσειῖε,
ἐστιτῆλδ᾽στσιλδσιρσαρ
μᾶκσι τιδ᾽λειῖσε
πιεῖστῶ συφφτε
ἔιικσιλσιμσι
τσῦσηδ᾽σι
σισσισμω
τ᾽εσ σιτσι, κειλ
τσιωσινγσεφτε
Δλσσηψυπτειλῖισε

has long featured in folklore. It is said that when Lucifer was cast out of heaven by the Archangel Michael he landed in a bramble bush and it is from this that the devil's association with the bramble stems. It is stated that come Michaelmas each year the devil goes about the land marking all the unpicked blackberries. The story of how he went about marking these berries varied depending on which part of the country you were from, in some areas he simply stood on them with a mucky hoof, in other areas he spat on them, while in other parts he just urinated on them. Whatever he did to them it wasn't healthy and it meant it was best not to eat them.

There is a little trouble in determining the date, the modern Michaelmas is 29 September while the old one is 10 October, and around where I live you have until early November before the witches urinate on the berries, yes it appears that the devil has subcontracted this work out. Of course today it is suggested that roundabout those dates the likelihood of the berries being riddled with horrible fungi increases, well that's what the powers that be would tell you.

So how do we get to the man-eating plant? My theory is simply this, that the combination of endless tangles of strong barbed runners and branches together with the tempting berries are part of a trap as insidious as the lobster pot trap with its easy-to-get-into but hard-to-get-out-of barbed branches.

You will often find that the best, biggest and tastiest berries will be just out of reach, just beyond your fingertips unless you stand on your tiptoes and lean just that little bit further. Now this where the location of the plant comes in handy because brambles seem to grow in some of the most unpromising places, the best plants around where I live hang temptingly over a disused railway bridge while others can be found on the site of an old military base with tangled mats of them concealing hidden drops. It is in uncompromising locations that this predatory plant works best. In best final destination tradition our luckless blackberry picker leans just a little bit too far forwards and falls base over apex into the heart of the bush becoming entangled in the barbed strands of the plant, add to this a hidden drop and our poor victim becomes stuck, and facing a slow death, all the while the bramble awaits for the victim to rot down adding nutrient to the area for its benefit.

You may think that this sounds a little farfetched but you would be surprised that it

is a scenario that appears to happen more times than you would care to believe. On the 2 September 2011 the *Daily Mail* contained a report of an 8 year old boy who lost his footing while blackberry picking resulting in him falling into a canal where sadly he drowned. If we substitute a steep-sided ditch for the canal we can easily see the feasibility of this hunting strategy. Other reports tell of cattle being entrapped in the grasping tendrils of the bramble unable to free themselves, they too meet an unlikely and unpleasant end.

Still not convinced? Well ask yourself how many times have you heard of a body being found in the undergrowth. Not all of them are identified as being suspicious deaths.

As a final word if you do go blackberry picking never go alone and always keep you feet firmly planted on the ground no matter how tempting that berry looks.

Blackberry entrapment in practice
In the following sketches I aim to display the blackberry bush in action. This is based on a plant at a nearby disused military site. For ease of displaying the plant in action I am showing it in cross section. The plant we are looking at is firmly rooted in a brick lined pit.

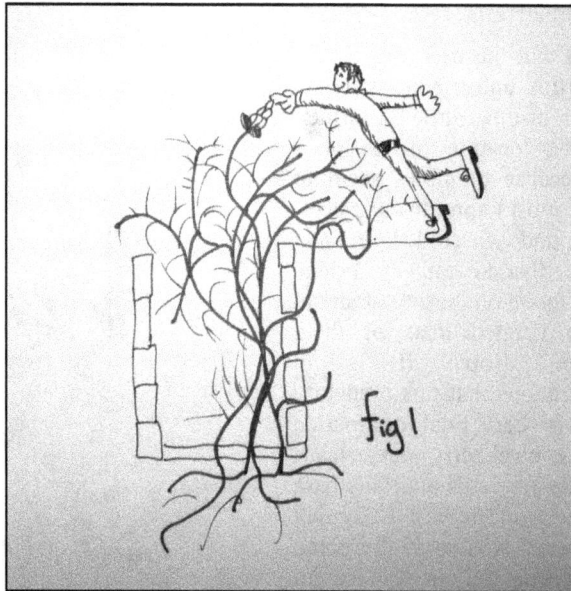

Fig.1
In this picture we can see our hapless blackberry picker stretching to reach the most tempting of berries. You can see that to reach the berry the picker is leaning over just a little bit too far. You can also see how dense is the mass of barbed branches that cover the hole in the ground.

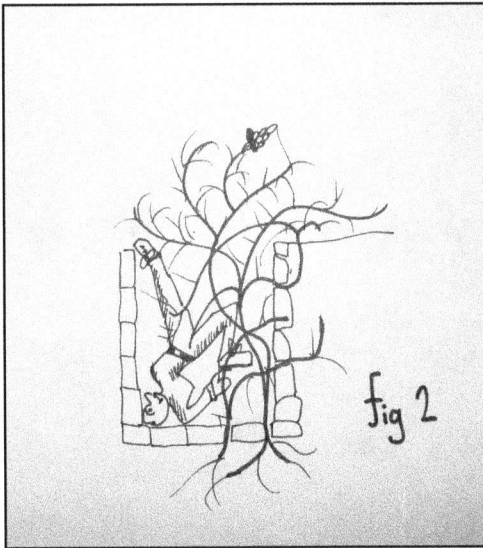

Fig.2
We now find that our hapless fruit picker has discovered the effects of gravity. Having fallen into the hidden hole headfirst he has landed badly with broken limbs and is completely entangled in the bush. Note that the tempting berry is still on show above the pit awaiting the next victim.

Fig.3
Here we see the blackberry bush next season with a new tempting berry blowing about in the breeze while below the remains of the last victim are a stark warning to the foolish fruit picker.

The killer in the garden:
A close up of a bramble showing the endless knot of barbed branches.

Sources

- *Daily Mail* 2011
- Golding, E. *The Country Book* (Ward, Lock & Co. 1937)
- Mackal, Roy *Searching for Hidden Animals* (Cadogan Books 1983)
- Mitchell, John & Rickard, Robert *Living Wonders* (Thames and Hudson 1982) Reader's Digest *Field Guide to the Wild Flowers of Britain* (Reader's Digest 1981)
- Reader's Digest *The Countryside Detective* (Reader's Digest 2002)
- Reader's Digest *The Gardening Year* (Reader's Digest)
- Reader's Digest *Wildlife on Your Doorstep* (Reader's Digest 1990)
- Sterry, Paul & Press, Bob *Wildflowers of Britain and Europe* (Selecta Books Ltd 2002)

Possible sighting of unknown ape species, Tibet 2009

Jonathan Jacobs

The photo was taken on the fourth day of an eight day tour of Tibet on September 18th 2009 and I have been diligently pursuing research into it ever since.

The photo data shows the picture was taken at 2.51am EST which was 2.51pm Chinese Standard time that day. I have noted that at present my camera's internal clock is running 9 minutes fast. So we can say more accurately 2.42am/2.42pm.

The tour was with a Chinese company, Access Tibet Tour. We had arranged for a good cultural and natural highlights tour running more or less southwards from our arrival point in Lhasa to Zhangmu and the Nepalese border. The tour came ¾ quarters of the way through a 6 month backpacking excursion across Asia.

On day four we departed Lhasa and our destination point for the night and following day was Shigatse. After experiencing the wondrous beauty of the super blue Yamdrok lake we set of for Karola glacier. As had become usual I would take the occasional photo of the inspiring scenery as we drove along.

What made me snap it? Was it the sense of something dark in the middle of all that grey, a cave perhaps? Certainly worthy of a few quick seconds of impulse. A zoom to maximum and shoot.

I don't know if it was that night or the following that I flicked through my photos, saw the dark image standing in the alcove and decided to zoom in for a better look. Astounded but mystified to see something that had to be a 'yeti', I accepted that I would look into it further.

After this I was quickly thrown back into the whirl of heady medieval Buddhism, high altitude and natural magic. It wasn't until the night of Day 6 where I found time to show Basang our Tibetan guide and Shudoy our Chinese driver the photo. As I zoomed in Shudoy was shocked,

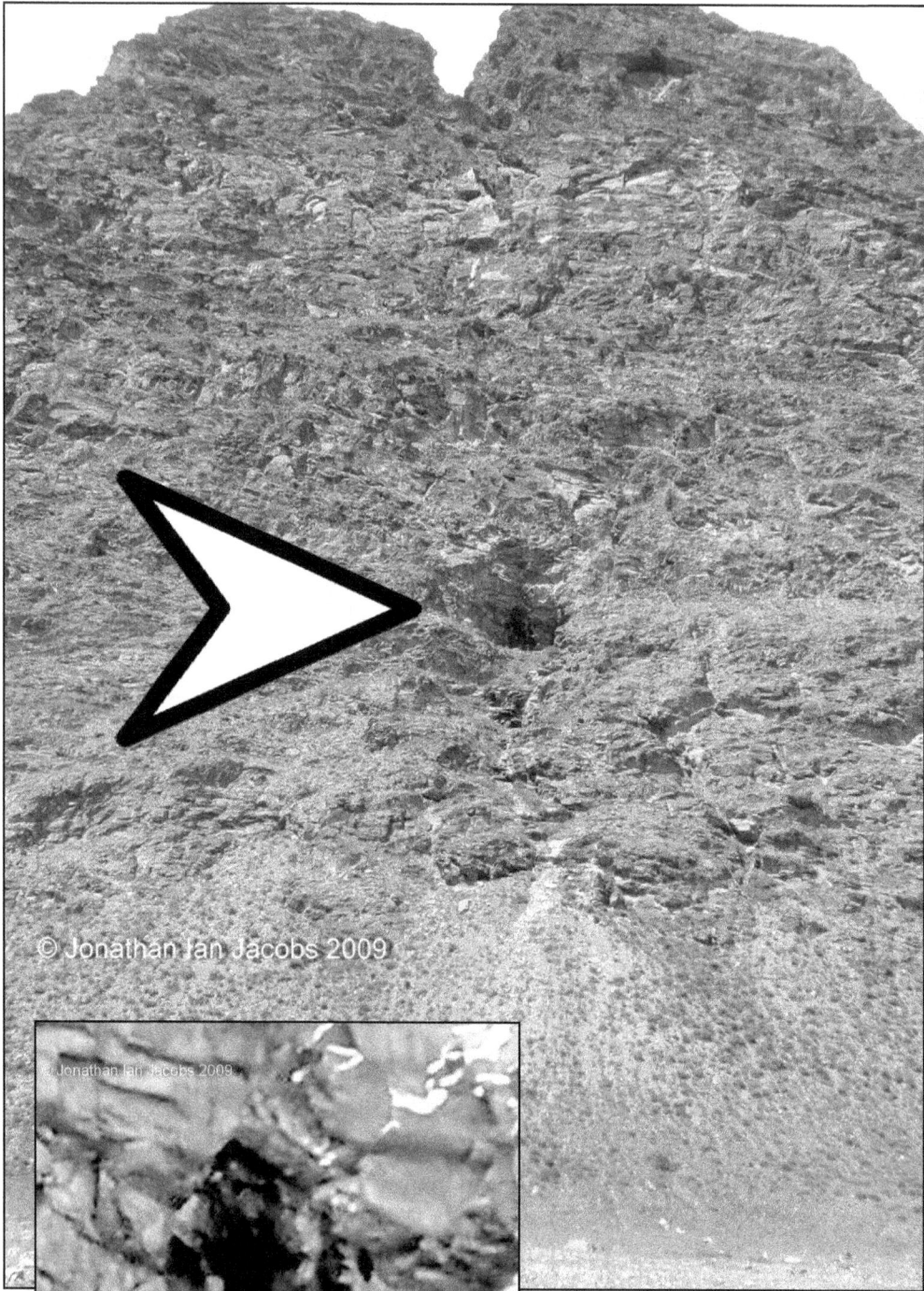

© Jonathan Ian Jacobs 2009

© Jonathan Ian Jacobs 2009

he didn't speak English but we had never communicated so much as we did in those few seconds. Basang was impressed and surprised but calmly said that it was probably a yeti, a creature before man. He'd never seen one, but it seems Tibetans accept the idea of the yeti quite naturally.

If the subject is animate however I would theorise this to be an evolutionary relative of the Gigantopithecus.

Analysis by John Traynor

Location: 28.991° N Date: 18-09-2009 Timezone: (GMT +8.0)
89.747° E

Sun position table:

Local Time (GMT +8.0)	Azimuth (deg. from N)	Altitude (deg.)	Shadow length (multiplyer)
07:50	87.524	RISE	n/a
08:00	88.738	1.780	32.179
08:15	90.559	5.059	11.297
08:30	92.391	8.336	6.824
08:45	94.246	11.610	4.867
09:00	96.138	14.875	3.765
09:15	98.081	18.129	3.054
09:30	100.089	21.366	2.556
09:45	102.180	24.583	2.186
10:00	104.373	27.774	1.899
10:15	106.689	30.933	1.669
10:30	109.155	34.053	1.480
10:45	111.799	37.124	1.321
11:00	114.656	40.137	1.186
11:15	117.766	43.078	1.069
11:30	121.176	45.932	0.968
11:45	124.939	48.680	0.879
12:00	129.116	51.297	0.801
12:15	133.772	53.754	0.733
12:30	138.974	56.016	0.674
12:45	144.776	58.039	0.624
13:00	151.215	59.776	0.583

Sun position chart:

Notes:

All angles (azimuth) relative to true north, and not magnetic north, which varies by location

Times are in the local timezone set (GMT +8)

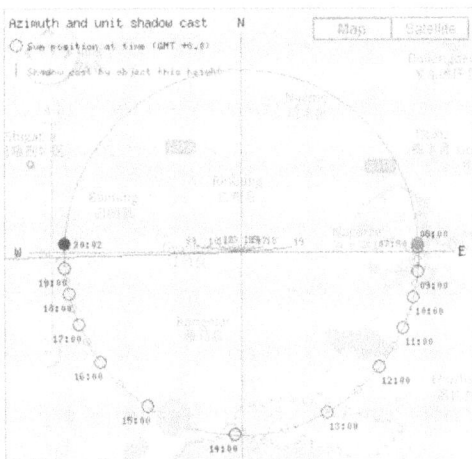

APPROXIMATE POSITION OF THE SUN @ 14.42HRS

13:15	158.279	61.174	0.550
13:30	165.896	62.181	0.528
13:45	173.916	62.753	0.515
14:00	182.122	62.862	0.513
14:15	190.259	62.504	0.520
14:30	198.088	61.695	0.539
14:45	205.424	60.474	0.566
15:00	212.158	58.888	0.604
15:15	218.255	56.992	0.650
15:30	223.730	54.835	0.705
15:45	228.632	52.463	0.768
16:00	233.024	49.916	0.842
16:15	236.971	47.225	0.925
16:30	240.537	44.417	1.021
16:45	243.779	41.513	1.130
17:00	246.746	38.531	1.256
17:15	249.483	35.484	1.403
17:30	252.025	32.384	1.577
17:45	254.405	29.241	1.786
18:00	256.649	26.062	2.045
18:15	258.781	22.854	2.373
18:30	260.822	19.623	2.805
18:45	262.789	16.374	3.403
19:00	264.698	13.110	4.294
19:15	266.563	9.837	5.767
19:30	268.399	6.558	8.699
19:45	270.217	3.275	17.475
20:02	272.393	SET	n/a

The original picture was focused in on the area of interest, up scaled, and sharpened, to identify the shape of the area of interest. Noted in the top left hand corner of the outline are the clear signs of light reflection suggesting the outlined figure was protruding from the shadow and into the light of the sun.

PHOTO FACTS: BACKGROUND.

The photograph was taken at 14.42hrs on 18th September 2009, at a point between Yamdrok Lake, heading North Westerly towards Shigatse, at an undefined point.

The photograph is believed to have been taken using the following equipment and settings:
Camera Maker:
Olympus Imaging Corp
Model: u1010, s1010
F-stop: f/3.5
Exposure time: 1/500 sec.
ISO speed: ISO-80
Exposure bias: 0 step
Focal length: 7mm
Max aperture: 3.44
Metering mode: Pattern No Flash
Digital Zoom: 1
EXIF version: 0221

It is also believed that the photograph was taken during motion along a road, as during enhancement considerable pixel loss through motion capture is experienced.

ANALYSIS PART 1:

In the first stage of analysis, the estimated positioning of the sun in relation to the location and time was tracked, and the data from this is displayed on page 1 of this report.

At the time this photo was taken, at 14.42 hours, the sun was in the south west of the area, with an estimated position of:

Azimuth (deg. from N)	Altitude (deg.)	Shadow length (multiplyer)
205.424	60.474	0.566

Figure 1 is an enhanced/sharpened copy of the original image, which has been altered to highlight the shadowing on the rocks. It is clear from this image that the angle of the shadows on the rocks suggest that the sun would have been high above, North West of the position of the photographer placing the location in an area furthest from Gyangze heading Northwesterly towards Shigatse.

Figure 1:

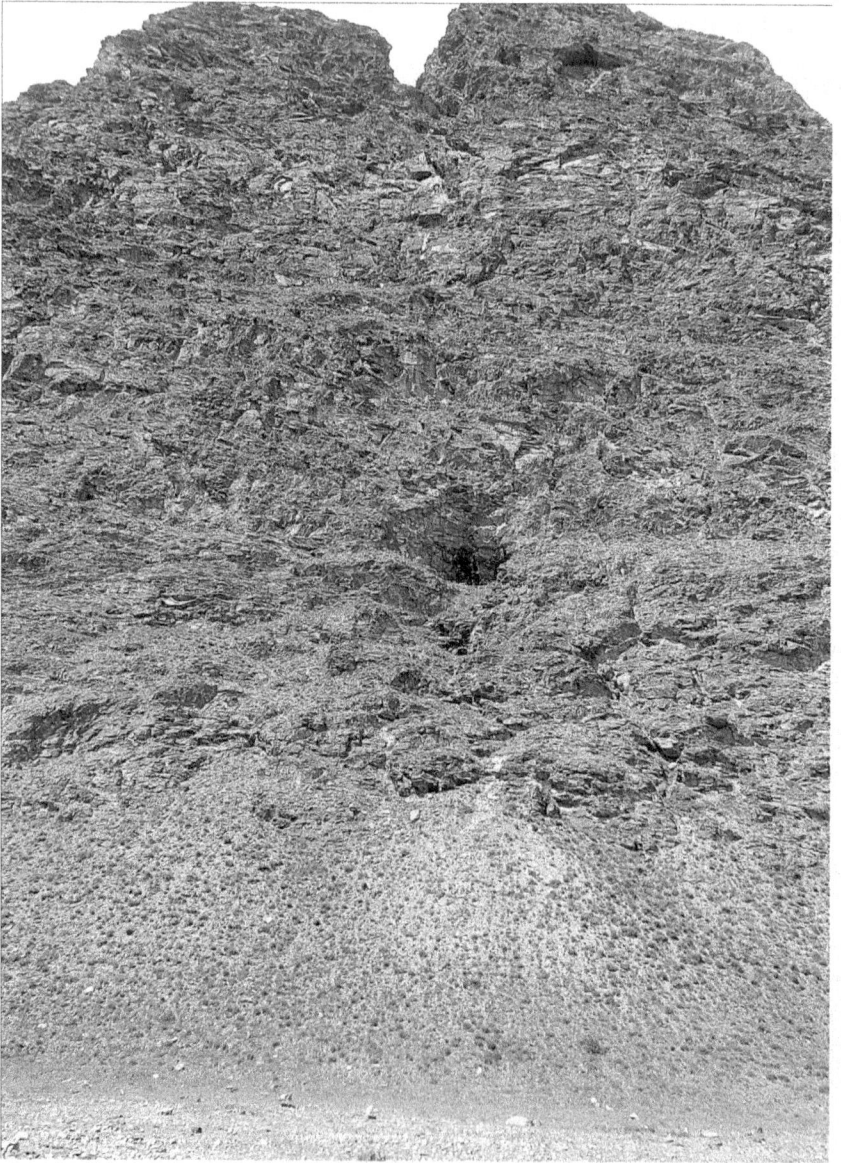

Considerable sharpening was applied to this copy, with a view to identifying the positioning of the sun at the time the photo was taken, partly eliminating the area of interest as simply a shadow.

Limitations of the picture analysis:

1) Motion Pixelation: It is acknowledged the picture was taken in motion, and as such, considerable pixelation exists in the enhancement/upscale of the photograph, with considerable pixel loss particularly visible when adding the effects.
2) The area of interest on the photograph is contained within an area which is less than 5% of the total photograph; The photograph was taken from a distance, so combined with the motion pixel loss above, significant loss of quality is noted when upscaling and focusing on this area of the photo.
3) The area of interest on the photograph is contained within a partially light exposed section of the hillside rocks, and much of the area of interest remains in the shadows, further adding to the darkness, and limitations detailed above. In order to alleviate these elements, considerable focus, exposure, hue, saturation and other levels had to be adjusted as per the pictures below.

Figure 2:

The original picture was focused in on the area of interest, up scaled, and sharpened, to identify the shape of the area of interest. Noted in the top left hand corner of the outline are the clear signs of light reflection suggesting the outlined figure was protruding from the shadow and into the light of the sun.

Figure 3:

Within this copy of the photograph, a heat emission effect was added in the form of Inverted data
from the photo, exposing areas of heat, and areas of cold as well as highlighting further areas of light
and darkness. The areas exposed to light are clearly lighter: The obvious exception being our outline

Figure 3:

Within this copy of the photograph, a heat emission effect was added in the form of Inverted data from the photo, exposing areas of heat, and areas of cold as well as highlighting further areas of light and darkness. The areas exposed to light are clearly lighter: The obvious exception being our outline of interest, which although already identified as being in a darkened area of shadow, comes up on this photograph as considerable light, further suggesting an association with heat either in the immediate surroundings or within the figure itself: This may be body heat, but cannot be verified through photography alone.

Figure 4:

Figure 4:

Upscale,Enhancement,Sharpening,Over Exposure, alteration of the saturation, contrast, brightness & hue of the picture further clearly defines the outline of a standing figure, with parts of the left hand body and face partially exposed in the sunlight protruding from the rocks above.

Figure 5:

Figure 5:
Within this image, the area of interest was cropped from the original, upscaled, sharpened, inverted & a black n white effect added, further sharpening, to define many of the primary features of this area of interest. What is clearly visible from this picture are what appear to be some facial features, including a considerable brow ridge above deep set eyes. The areas of the forehead are clearly highlighted here also particularly the features which appear to be in partial sunlight: Further analysis took place on the forehead the result of which can be found below.

Also clearly defined within the bottom left hand section of this picture is part of what looks like the hand or foot, with what appears to be 2 claws clearly visible and defined from the background.

Figure 6:

Figure 6:
I felt it was important to focus in (as much as I was able to) on the forehead area, and carried out some focus, upscale, sharpen, and over exposure effects, followed by an alteration of the Hue/Sat Levels. What are clearly defined within this up scaled copy of the image are sections of green areas which appear throughout the original image if up scaled with the above named effects. Having carried out a colour survey, the shade of green ranges from HEX Value #4e5206 at its darkest points, through to HEX Value #fff1bd at its lightest.

The colour variation is evident like this:

Figure 7:

What this variation/gradient indicates is that the green within this picture is a variation of the natural colour of the rocks in the background as there is a direct correlation between the darkest and lightest parts of this environment as per the gradient above.

Given the position of the sun when this picture was taken, coupled with the light environment of the area of interest, I believe this patchy green area to be a motion blur which is evident deep within the pixels of the photo, further adding evidence to suggest that this photo was taken from a distance and whilst in motion. Given the variation and directional motion of the pixelation which exists within this colour gradient I would assume that the photo would have been taken from the right hand side of the vehicle whilst travelling in a westerly direction. This is an assumption which I cannot verify based simply on the evidence this photo presents.
In addition to this, the reflection on the 'forehead' is also interesting. Note the intensity of a direct reflection on the left hand side of the 'forehead', compared to the right. The intense direct reflection on the left fades out into the middle of the 'forehead' before re-emerging on the right hand side, blending in to the gradient motion blur of the background environment. I would suggest that whatever the entity is, it appears that this photo was taken at a split second when the 'head' was turning slightly to the left and up, giving motion blur from the background, direct reflection on the left hand frontal lobe of the 'forehead', and an emerging reflection on the right hand side. This is further evidenced by the presence of an emerging direct light reflection on the primary features of the face, further suggesting the head was turning at the time this photo was taken, albeit slightly. The movement of the 'head' is evidenced through the levels of reflection, coupled with the motion blur and distance.

Figure 7:
Within this image, I applied an upscale, enhancement, sharpen, exposure,smart sharpen, and pixelate effect, to draw some of the facial features from the 'head'. This is as clear as this image could get, due to the distance light and motion blur. The head shape is defined, as is the brow ridge with the deepened eye set.

CONCLUSION:
Much time has been spent within this analysis determining light levels, and the origin of direct reflection in establishing the nature of the highlighted figure.

I note that several contour lines of the rocks within the background of the figure continue unblocked from view through the figure adding doubt as to whether this figure is of a real being or indeed the figure of an ape like subject.

What is clear from the number of modes contained within this report is that the figure clearly stands out from the background when exposure is highlighted and contrast, saturation and hue are modified.

In addition, the outline of the figure has been analysed with visible direct reflection points noted, and the correlation between these, the position of the sun at the moment the photo was taken, and motion blur all noted within their own contexts within the picture.

Unfortunately, due to the motion, distance and size of the area of interest within the photo, it has not been possible to definitively establish the exact details of either the eyes, or hands/feet, the only exception to this being the obvious 2 claw like shapes highlighted within numerous modes of the photo.

I believe the results to be entirely subjective and down to individual interpretation, though the outline, reflection and features of an indefinite figure cannot be argued or contested easily.

British Snake Catchers and Eccentrics

Richard Muirhead

Once again I return to the theme of snakes (this is getting worrying, was I a snake in my last life? I must have been a very good little snake to re-incarnate as a human and make that species jump so rapidly) having written about 'Some Strange Snake Stories' in the 1998 CFZ Yearbook and 'Flying Snakes and Jumping Snakes - a Worldwide Survey' in the 2010 Yearbook.

British snake catchers are part of a long tradition of eccentrics from rural areas including rat catchers, Morris Men, radical preachers, rag-and-bone persons, metal detecting enthusiasts as I once was, culminating in cryptozoologists and Forteans of course. Snake catchers go back as far, at least, as the end of the 17th Century.

The Philosophical Transactions, Medical essays and observations, (1734), p. 324 ff contained 'Some observations on a man and woman bit by vipers' - Dr Atwell (of the Royal Society.) :

> "July 8:Two young Pigeons were bit, the one had Viper-Oil applied immediately, but it sickened and died in four Hours: The other had Oil-Olive applied, and recovered perfectly , the Flesh beginning to return to its natural Colour in about an Hour`s Time.
>
> July 17: The Woman was bit in the publick Hall of *Brazen-Nose College* in the Presence of Dr. *Frampton* Dr *Frewen* and several other Physicians. It had been suspected , that they played some tricks with their Vipers, and made them spend their Rage and Venom before hand: To obviate which, a Physician of the Company had provided some fresh Vipers which he had caught himself a Day or two before, and kept in his own Custody till that Time. The Woman was bit twice by one of these and received three Wounds, one in the Thumb, and two in the Fore-Finger; her Hand was soon swollen and spotted and her Finger turned black..."[1] And it continues along this rather gruesome vein describing the affect of the venom and its cure.

Some years later, William Ellis (1749) pp 105-7 in his *A Compleat System of Experienced*

Improvements, Made On Sheep, Grass-Lambs, and House-Lambs: Or, the Country Gentleman's and the Shepherd's Sure Guide... writes of adders or vipers biting sheep or lambs:

Having described two "receipts" or methods of curing bitten sheep or lambs: *A Third Receipt for the bite of an Adder or Viper, or Slowworm* says –

> "If a Sheep`s Udder is bit by any of these, you may make use of the following Remedy : Take oil of Scorpions and Vinegar, with Plantane Leaves and Bole Armoniak, made thick like a Salve, and anoint the bitten Part with it three times a Day.

> *A fourth Recipt –*" TAKE Sanguis Draconis (what have we here? This literally translates as "blood of a dragon. A web site describes it as a herb known to the Chinese which, when applied externally, stops bleeding.) a little Barley-Meal , and the White of an Egg, which beat together , and lay it Plaister-Wife on the sore, renewing it once in Twelve Hours, say Several Authors- But I am of Opinion, that the Olive Oil drove into the Sheep`s Udder by a Hot Iron, and Oil given inwardly, is the best Remedy except Viper-Fat: For even a hot Iron alone has such an attractive Quality, that if it is immediately applied hot enough to such a Bite , it is a Remedy of itself – But for a Bite of a Viper or Slow-worm, I could enumerate and give an Account of several other Remedies, without being beholden to Ancient authors for what they have wrote on this Subject: However, at present, I must forbear to do it, because it will not be agreeable to my Purpose in the Work [2]

According to the *Medical Repository*, (1805) p. 242 `On the effects of oil in cases of bites of serpents`:

> "In great cities, particularly in London, a number of persons procure their livelihood by catching vipers. They are employed by chemists, apothecaries, etc."[3]

The poet John Clare (b. July 13th 1793 - d. May 20th 1864) wrote an essay called simply `Snakes`, first composed in 1824 but not published until 1951; a part of which is included below. [transcribed in the style of the original characteristic punctuation-less manner and distinctive spelling used by Clare at the time - Richard]

Snakes

> I do not know how to class the venomous animals further then by the vulgar notion of putting toads common snakes black snakes calld by the Peasantry Vipers Newts (often calld eatherns) and a nimble scaly looking newt-like thing about the heaths calld Swifts by the furze calld Swifts by the furze kidders and cow keepers all these we posses in troublsome quantitys all of which is reckond poisonous by the common people tho a many daring people has provd that the common snake is not...yet still they are calld poisonous and dreaded by many people and I myself cannot divest my feelings of their first impressions tho I have been convinced to the contrary...when they first leave their shells they are no thicker then a worsted needle or bodkin they nimble about after the old snakes and if they are in danger the old ones open their mouths and the young disappear down their throats in a moment till the danger is over and then they come out and run about as usual I have not seen this myself but I am as certain

of it as if I had because I have heard it told so often by those that did...people talk about the Watersnake but I cannot believe otherwise then that the water snake and land snake are one tho I have killd snakes by the water in meadows of a different and more deep color then those I have found in the fields the water snake will swallow very large frogs I have often known them to be ripd out of their bellys by those who have skind the snakes to wear the skin round their hats which is reckond as a charm against the headach and is often tryd but with what success I am not able to say... the black snake or Viper a very small one about a foot long and not often thicker then ones little finger is very scarce here and venomous I believe the fens have none they seem to inhabit high land...I have seen three of these black snakes they are very quick eyd looking things with a fang darting out like common ones their heads are shorter and much flatter then the large snake and their colors are more deep and bright their backs are black and their bellys bright yellow interspersed with scaly bars of blackish hues – I have heard some people affirm that even these are not venom [ous] and that people who suppose themselves bitten by them mistake sudden yumours falling in their limbs for a bite. I believe this is the Doctors opinion with us – all I can say is that I never was harmed by them[...][4]

I also have the following from the *Farmers Journal* of 1834 – `Viper catcher in England.`

"I have read of a man in England who made his living by killing adders and selling their fat, which he used for its medicinal virtues. His dog that had been taught to find the sluggish adders, was sometimes bitten by them, but the poisonous effect was always quickly removed by rubbing the wound with adders fat. This is the only evidence that has reached me of the adder being poisonous, and it serves as well, to prove that their poison is followed by very little pain or danger." [5]

In 1848 *The Plough the Loom and the Anvil* (1848 p.393) published the following about a Brighton viper catcher:

" The viper catcher whom I met with near Brighton assured me that he had frequently seen the young vipers take refuge in the inside of their mother by running into her mouth, which she opens for that purpose, when danger is apprehended. He also asserted that they are produced alive, the ova being hatched in the inside of the mother, from which they probably creep, as they must do at a more advanced state after they have made it their place of refuge. He said that, by letting vipers bite a piece of rag, and then suddenly snatching it from their mouth, he easily extracted the fangs, and that he then frequently put the animals between his shirt and skin, and brought them away alive." [5]

There is also an interesting report of a certain Jonathan Hulme in the *Manchester Guardian*`s Local Notes and Queries on October 12th, 1874:

JONATHAN HULME – The following account of a local eccentric may perhaps have its interest for your readers. It is given in Hulbert`s " Memoirs" , p.99 : - " Poor Jonathan [Hulme] imagined he had a commission from heaven to destroy all the serpents and vipers who had made their dens in this desolate region Chat Moss. He would have it the serpents God sent among the children of Israel, that destroyed such numbers, were of the same species as those which infested Chat

Moss. No actual derangement of intellectual was perceptible in Jonathan, excepting on the subject of viper killing, hence he assumed the title of `General Viper Killer`. Often has he amused all our family with the stories of the ferocity of these serpents, the dangers he escaped, and the victories he had won. His habits and attire were all in perfect keeping; he had a girdle of viper skins around his loins and a band of the same round his hat; in each hand he carried a staff or spear armed at the points with sharp knives and a clasp for catching and keeping hold of the serpents when he did not desire immediately to destroy them. " – W.R.C. [6]

Also Thomas Hardy , in *The Return of The Native*, 1878, mentions them as follows:

"I know what it is, cried Sam. " She has been stung by an adder!"

"Yes, said Clym instantly.I remember when I was a child seeing such a bite. O, my poor mother!"

"It was my father who was bit, " said Sam. " And there's only one way to cure it. You must rub the place with the fat of other adders, and the only way to get that is by frying them. That's what they did for him."

"Tis an old remedy, said Clym distrustfully, "and I have doubts about it. But we can do nothing else till the doctor comes."

"Tis a sure cure" said Olly Dowden, with emphasis. "I've used it when I used to go out nursing."

"Then we must pray for daylight, to catch them" said Clym gloomily

" I will see what I can do," said Sam

He took a green hazel which he had used as a walking stick, split it at the end, inserted a small pebble, and with the lantern in his hand went out into the heath. Clym had by this time lit a small fire, and despatched Susan Nunsuch for a frying pan. Before she had returned Sam came in with three adders, one briskly coiling and uncoiling in the cleft of the stick, and the other two hanging dead across it.

" I have only been able to get one alive and fresh as he ought to be," said Sam. " The limp ones are two I killed today at work; but as they don't die till the sun goes down they can't be very stale meat." [7]

Apparently the infant Hardy was found with "a large snake curled up on his breast in his cradle." [8] This was according to Florence Hardy, his wife from 1914 and biographer after his death in 1928.

Perhaps the most famous of British snake catchers is Brusher Mills (b. March 19th 1840 - d. July 1st 1905). He lived and worked as a hermit in the New Forest and it has been said he

---

* Perhaps elvers (baby eels)? Ed.

killed 29,233 grass snakes in his life time and 3,854 adders and (of interest from a cryptozoologist's point of view) 200 smaller things he called `Levers`, "probably a local name" [9] whatever they might be *. Legend has it that he once emptied a bag of snakes onto the floor of his favourite pub *The Railway Inn* in order to get to the bar. [10]

By the time of his death he had become something of a tourist attraction and a celebrity. It is worth quoting extensively from a pamphlet *Snake Catchers of the New Forest*, by David Stagg who remarks:

> "No doubt there have always been snake catchers to meet the demand for snakes, and at the same time to clear areas infested with snakes. In the royal forest there were officers known as Verminers, and probably their responsibilities would have included snakes along with mice, rabbits, and other pests. At the end of the 17th century this office became extinct in the New Forest, but no doubt the work was continued by private individuals such as Brusher Mills....Brusher Mills at the time of his death was described as one of the chief personalities in the New Forest, and one and one who had attracted a unique public eminence. This had resulted from his colourful character, his unusual life style, and his bizarre occupation, but perhaps of most importance he lived at a time when the New Forest was rapidly becoming a major tourist centre. This created an immediate demand for guide books, post cards, and local colour, and this demand was fully met in the person of Brusher Mills. At least as early as 1887 he was described in a local newspaper article, and was then said to have been catching snakes for six years, and these he sold to the London Zoological Gardens for one shilling each. His equipment consisted of a stick about 4ft in length and with a forked end which he used to pin down a snake, and a pair of very long scissors the end of which were flat and blunt. These he carried in the buttonholes of his coat and used to pick up the snake from behind the head. At this time he already acted as a guide to visitors, and in later years he spent much time in showing his snakes and catching equipment to tourists, also in demonstrating his skill at recapturing an "escaped" snake, for which he would be suitably rewarded." [11]

In 1998 I wrote:

> "In 1887 an unnamed newspaper in the Museum at Devizes, Wiltshire, reported that the then famous New Forest snake captor "Brusher Mills" asserted that the red adders never grew any larger than their usual small size and were equally as venomous as the larger brown species." [12]

The knowledge of Hampshire reptiles at the time was displayed by the Rev J.E. Kelsall who in 1897 mentioned in 'The Reptiles of Hampshire and the Isle of Wight' in: *Papers and Proceedings of the Hampshire Field Club,* vol 3, pp 267-73, that the grass snake, adder and smooth snake existed in Hampshire and that Canon [Charles] Kingsley had said:

> "But I have only time to point out to you a few curious facts with regard to reptiles, which should be specially interesting to a Hampshire bio-geologist. You know, of course, that in Ireland there are no reptiles, save the little common lizard, *Lacerta agilis*, and a few frogs on the mountain-tops—how they got there I cannot conceive. And you will, of course, guess, and rightly, that the reason of the absence of reptiles is: that Ireland was parted off from England before the creatures, which

certainly spread from southern and warmer climates, had time to get there. You know, of course, that we have a few reptiles in England. But you may not be aware that, as soon as you cross the Channel, you find many more species of reptiles than here, as well as those which you find here. The magnificent green lizard which rattles about like a rabbit in a French forest, is never found here; simply because it had not worked northward till after the Channel was formed. But there are three reptiles peculiar to this part of England which should be most interesting to a Hampshire zoologist. The one is the sand lizard (*L. stirpium*), found on Bourne-heath, and, I suspect, in the South Hampshire moors likewise—a North European and French species. Another, the *Coronella lævis*, a harmless French and Austrian snake, which has been found about me, in North Hants and South Berks, now about fifteen or twenty times. I have had three specimens from my own parish. I believe it not to be uncommon; and most probably to be found, by those who will look, both in the New Forest and Woolmer. The third is the Natterjack, or running toad (*Bufo rubeta*), a most beautifully-spotted animal, with a yellow stripe down his back, which is common with us at Eversley, and common also in many moorlands of Hants and Surrey; and, according to Fleming, on heaths near London, and as far north-east as Lincolnshire; in which case it will belong to the Germanic fauna. Now, here again we have cases of animals which have just been able to get hither before the severance of England and France; and which, not being reinforced from the rear, have been forced to stop, in small and probably decreasing colonies, on the spots nearest the coast which were fit for them." [13] (Charles Kingsley, On Bio-Geology, 1871)

In 1900, a newspaper article gave an indication of Brusher's activity at the turn of the century:

Hampshire Advertiser March 11th 1900 p.8
LYNDHURST

"Brusher" Mills the "New Forest snake and adder catcher" up to the present time has the following remarkable score: Snakes 29,025; crown snakes, 198; adders 3,848; making a total of 33.071 reptiles. Brusher was out this week, and received a bite in the right thumb, but on Wednesday he had it set right, and was able to resume his occupation. [14]

On May 5th 1904 p.2 the *Portsmouth Evening News* contained this story:

A GREAT SNAKE-CATCHER
IN THE NEW FOREST

"Interviewing Brusher Mills, the snake catcher of the New Forest, a writer in the May "Windsor" says: No tramp would have the steady walk , the upright bearing, or the steady look of the old man, or carry the cleft stick so well suited for stopping a snake, the forcep-like scissors for picking it up, the tin to receive it. He passed me a civil greeting, and I signed to him to come to the porch...

"Are you Brusher Mills? I said." "That's my name," he answered.

He speaks very indistinctly, for he has a cleft palate, attributed by some of the forest-folk to the machinations of the snakes, whose enemy he has been for so many years. "The snakes tried to dumbfound him" said an old woman, speaking to

Brusher Mills` REF:ZDM92/Z/QA9
Reproduced with permission of Hampshire Record Office.

me of Brusher Mills a few days after my first meeting with him – "tried to dumbfound him they did, so he couldn't say what he was doing on people's land, and he would be warned off for trespassing. But he kept enough speech to speak, an' now he's mortal hard on every snake he meets.They're sorry by now they could not close his mouth altogether." [15]

By October 14th 1904 Brusher had been dead just over 5 months, a newspaper reported the following. *Exeter and Plymouth Gazette* October 14th 1905 p.5 :

"The grave of "Brusher" Mills, the New Forest snake catcher, in the parish churchyard of Brockenhurst, is to be marked by a memorial stone, to be erected by public subscription. " [16]

In 1907 a writer to the *Countryside* journal, S.Yeates ,of Cowfold wrote the following:

" ...At the same time there was an old man who was called "Adder-catching Jack," who did nothing else that I know of but catch adders to sell to chemists for the same purpose. It may sound incredible, but he had a jacket made entirely of adder's skins sewn together, and it was perfectly watertight. I forgot to mention that the red rag on the rod served a double purpose, for when the adder seized it this gave the operator time and opportunity to peg the adder to the ground; and as my father told me the tooth of the adder is encased with a thin skin which is discharged in whatever it bites, the venom is absorbed by the rag, and the viper is rendered almost harmless for a time; and he said he would not be afraid to handle it, though I do not know that I ever saw him do so. – S.YEATES,Cowfold. [17]

In the *Countryside* on October 20th 1907 p. 351 the Rev. T. Wolseley Lewis wrote:

"Adder's Oil" –With reference to the opinion that adder's oil may be an antidote for adder's poison, the following information may be of interest to those reader's of THE COUNTRY- SIDE who are not already familiar with it. In Jerem viii 22, "No Balm in Gilead" was rendered "No Treacle" in an old version of the Bible; and the history of this latter word is very interesting and perhaps not generally known, and it illustrates the popular idea about adder's oil , or more probably it is the foundation of it. The world " treacle" is derived from the Greek word "therion", which meant first a wild beast any knd, and then more especially one which had a venomous bite, and by many Greek writers it was used to denote a viper specifically... (Rev.) T. WOLSELEY LEWIS, Eirianva, North Malvern. [18]

On May 19th 1922 the *Western Morning News* p. 6 published the following:

SNAKE CATCHING AS A HOBBY
PLYMOUTH MAN'S 60 YEARS' EXPERIENCES

The recent correspondence concerning vipers has brought to light the fact that in Beaumont-road, Plymouth, resides a gentleman who has spent the spare moments of a long life to the pursuit of these reptiles. This " champion snake catcher" is 78 years of age, and recently informed our representative

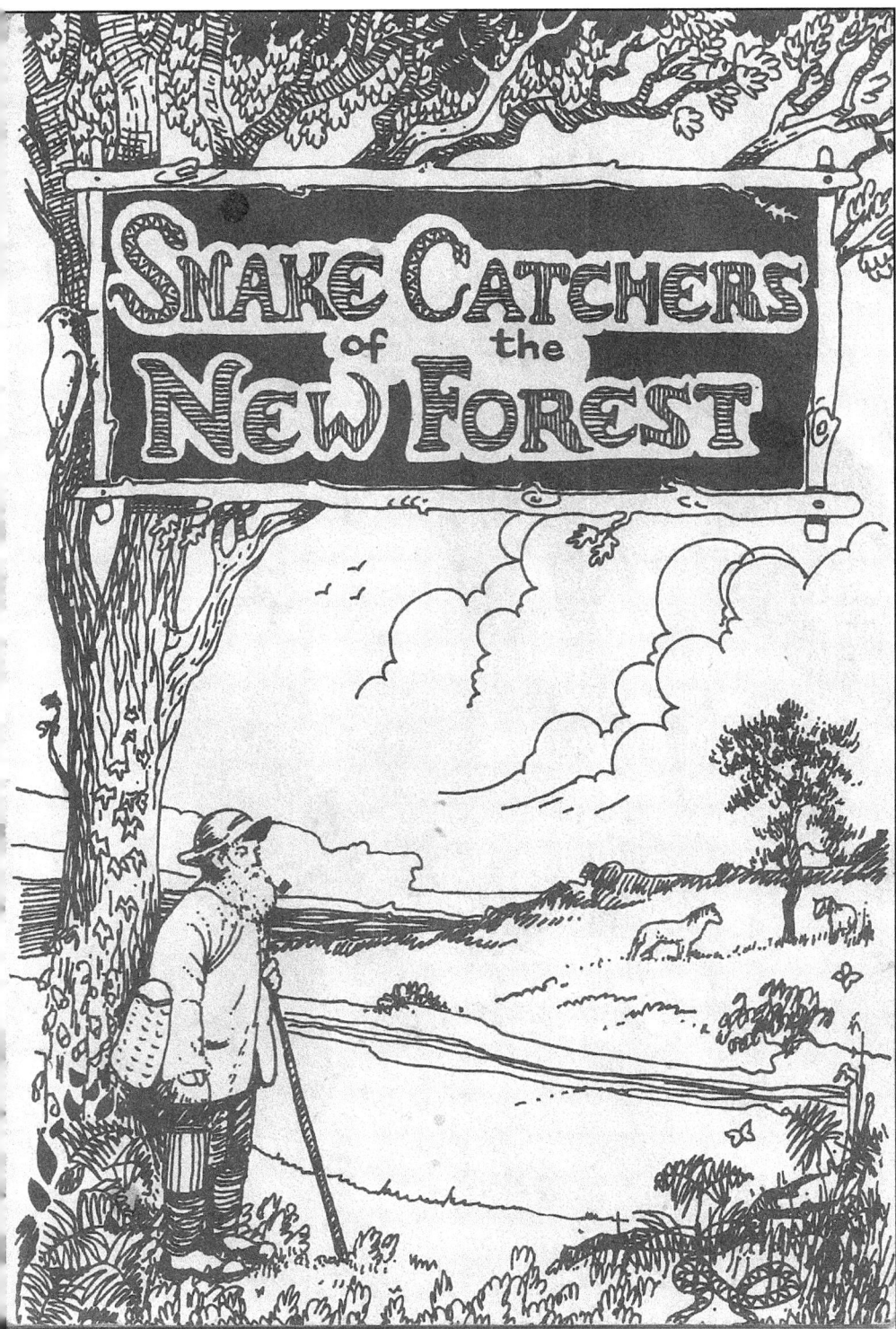

Author David Stagg, published 1983 by New Forest Association

Plymouth Snake Catcher. *Western Morning News* May 19th 1922. Reproduced with permission of British Newspaper Archive.

that he had engaged in this hobby since the age of 18.

" There is a great deal of nonsense talked – and written – about snakes," said Mr H. B. Hearder. " I used to supply the late Professor Huxley with live specimens for his experimental work. After a short time he wrote imploring me not to send them to his private address. Reading between the lines I guessed the secret of this request. His wife had objected!...The common or grass snake is absolutely harmless. It possesses one exceedingly objectionable habit, that of ejecting an odour like garlic when garlic when alarmed. Wash them, and during a long captivity they will never repeat this unpleasant trick unless they injure themselves in some manner. They grow to a length of three to four feet."

THE DANGEROUS VIPER

The viper is Mr Hearder`s speciality, and he has caught hundreds on Dartmoor, in the manner illustrated in the photograph...Mr Hearder...demonstrated the wonderful mechanism of the viper`s jaw in which rests the poison fangs, which are covered by a sheath when not in action. The viper will attack a man if provoked, and the snake catcher has experienced some alarming sensations on some of his exploits. As an example of the deadliness of this poison he quoted an experiment which he made:

A mouse was placed near a viper and, being bitten, died within 30 seconds of the attack. " Great swelling usually accompanies such a bite upon the human body" concluded Mr Hearder. " The only remedy is to suck the wound at once." [19]

In early September 2012 fellow Fortean Bob Skinner passed the following information on to me from his native Surrey: "One member of the [local village History Group] , a local Hale man (Roy Poole) now in his eighties, mentioned the following which he remembered:

Daniel Usher, a Hale man, lived in the Fleet Road, in the house next to where the Shomraat Indian restaurant is today, opposite the Hale School.

Mr Usher was in his 90s when Roy was in his early teens in the late 1930s. Apparently for many years from before the 2nd World War and during it, Daniel Usher used to go every Sunday onto the local heath land common at Upper Hale (near Farnham) and catch adders , using a forked stick. These he put in a bag, and then on his bicycle he used to take them to a local hospital in Farnham (on the site of the workhouse) where (Roy said) they were used to get serum [for medical treatment.] [20]

Many years later, in 1943 in fact, *The Field* described Brusher in the following manner:

" Readers may not have known that he was called ` Brusher` not as a man of the heath , but because he had once been employed as a sweep or `brush` at Lyndhurst cricket ground, and that he was not a Forest man by birth but, like many New Forest residents, had moved there from the town in early middle – age.

In true life he can be identified as Henry Mills, the son of Thomas and Ann Mills. His father came from Mottisfont and was not a forester. Henry was born on the 19th March 1840 and was one of a family of eight or nine children, and in addition there were five children from a previous marriage...It was not until the summer of 1884 that Brusher took to sleeping in a disused charcoal burner` hut where, on the evidence of his nephew, he remained for 19 years and 4 months until the hut was burned in the autumn of 1903....Some mystery surrounds the burning of the hut, but from official papers it appears that this did not take place until some time after Brusher had been told that he must leave. It was popularly supposed that this was done to prevent any possibility that Brusher might acquire squatter's rights to the hut, but the true reason was probably that the authorities were concerned about his deteriorating health... Another favoured habitat was the Railway Inn at Brockenhurst...Brusher was described as a short, stocky figure, about 5ft. 3 ins in height, with bushy brows, a forked beard, and a face the colour of mahogany. Due to a cleft palate his speech was indistinct, but those who could understand him said that when aroused his vocabulary was both extensive and profound. His outer dress consisted of one or two overcoats, creased trousers and gaiters, and heavy boots, together with a faded black felt hat and a neckerchief. His body was hung about with sacks and bags, and his pockets were stuffed with tins, boxes, and bottles

He placed great faith in the curative properties of adder's fat, and carried in the pocket a bottle containing some quarter of a pint, and which he claimed to be worth thirty shillings. After a snake bite he would bleed the wound, and then apply adder's fat...On the afternoon of Saturday 1st July 1905, Brusher had been in the Railway Inn drinking " two or three pennyworths of rum," this being his usual drink. He appeared to be in his usual state of health, but his nephew did comment that the previous day he had been offered a glass of beer by Brusher, a thing that had never happened before. After his drink Mills had eaten some bread and pickles, and then left the bar. When he failed to return a search was made and his body discovered in an outhouse.

The doctor who was called diagnosed that Brusher had died from valvular heart disease , and did not consider that a post mortem was necessary as he had been in failing health for the previous year, and only some months before had nearly died from heart failure while in the Rose and Crown. He said that it was a common thing for sufferers from heart disease to die after an indigestible meal such as Mills had just eaten.

The funeral was simple in the extreme, the mourners comprising a sister of the deceased, Mrs Finch, who had been found by the constable, George Abbott his nephew, William Perkins the landlord of the Railway Inn, and the village constable who had made the funeral arrangements. Subsequently a memorial stone was erected by public subscription." [21]

The following quote is from a web site, `The Snake Catcher (Brusher Mills)` devoted to the life of the snake catcher:

On the subject of the New Forest's snakes, its Gypsies believed that a good tonic could be made up from snake flesh and that a snake skin hung at the door of caravan could ward off the effects of the evil eye. But the Romany cure for rheumatism differed from "Brusher's", being more magical than practical. According to Wendy Boase's Folklore of Hampshire and the Isle of Wight , a Gypsy wanting to avoid such aches would carry the skin of a frog or an eel. [22]

A somewhat belated obituary of Brusher published in *The Country-Side* magazine of August 5th 1905 commented:

" Not with the magic flute of Pan, or Orpheus` lyre does he lure the snakes from their thickets, but pursues his somewhat dangerous craft armed only with a stout hazel stick with a forked tip, and a large pair of iron pincers like the inquisitors of old...His largest catch was 160 adders in a month, and the average bag in two years was five thousand serpents, of which one thousand were adders...He is a handsome old man, looking rather like Rip van Winkle tidied up, with his long beard and thick eyebrows." [23]

Another man who caught snakes, following on from Mills was George Wateridge (1869-1948.) He was born at "The Orchard" at Lyndhurst. Again, according to Stagg,

"In his early life he worked at Wilverley Park nearby, and subsequently lived in a cottage at Pond Head. As a young man he was well acquainted with Brusher Mills and no doubt sometimes joined him in snake catching. There is a family tradition that on the eve of his marriage George spent the night in Brusher`s hut, no doubt this being a sequel to an evening of celebration at the Crown and Stirrup. After Brusher`s death, George Wateridge inherited the tongs and forked stick, and took over the supply of snakes to the London Zoo and also to the Zoological Society of Scotland. When captured the snakes, to the terror of his wife, were in his garden at the Pond Head in an old dustbin with a little hay in the bottom, and when a sufficient number was available would be dispatched by train from Lyndhurst Road station. The carrying of the snakes to the station was a task performed by his eleven children of whom Mrs. Mary Broomfield is the youngest...In addition to snakes he would receive orders for lizards, slow-worms, frogs, and toads, all of which would be supplied. Also he became well known as an expert entomologist and sent butterflies and moths all over the country, these being taken by the sugaring of trees and the use of powerful acetylene lamps. Apart from snake catching being a seasonal occupation, George Wateridge never regarded the work as being more than part-time, and he was very much the typical forester and able to turn his hand to any other jobs which were available, but these would be interrupted if he received an order for snakes or a request to badly infested area...The tongs and forked stick of Brusher Mills then passed to Douglas Bessant, the son of Ben Bessant a forest keeper, who had become interested in snakes while on war service in Burma, and who for some time continued to supply Zoos and various educational establishments with snakes. However he was soon to emigrate so ending so ending the long tradition of snake catchers in the New Forest." [24]

[About the late 1940s-Richard]

Kelsall (see above) also said:

" I have no space to deal with the great controversy as to whether Vipers swallow their young . I fully believe that they do. I can say exactly what Gilbert White said in 1768, that " several intelligent folks assure me" that they have seen it. Among them are the Rev. G.M.A. Hewett, M.A. ., Assistant Master of Winchester College, and Mr Charles Crouch jun., a respectable parishioner of my own. Brusher Mills, the well-known adder catcher of Lyndhurst, declares he has often see it. Miss Hopley, authoress of the little work I have quoted, and a well-known student of Reptile life, fully believes it to be true, and remarks that the habit is known to exist among various lizards and fishes. Some persons are, however, incredulous, and Mr Tegetmeyer, of the " Field" Newspaper, offers £5 for a

specimen with its young in its stomach." (25)

In Cheshire there has been a long history of the use of adder oil for various purposes. There are many Adders Moss' and Adder`s Woods`, including in Cheshire. Here is an extract from *Hadsfield's Macclesfield Memories* by C. Hadfield. Hadsfield's is a very long established chemists in Macclesfield.

ADDERS MOSS

Andrew Allen in his splendid article in *The Field* in 1979 described the reputed magical properties of the adder.

These properties did not die out in England until the dawn of the twentieth century. I have in my possession the containers from which Hadsfields dispensed oil of adders and oil of scorpions, still in use in 1910.

The adders were dropped alive into boiling olive oil and the temperature maintained until the flesh fell from the bones.

Adders were prolific in parts of England and there are many Adders` Moss` and Adders` Woods` but so widely were they used in medicine in the Middle Ages that they had to be imported by the barrel from Italy and the Middle East.

There were several areas around Macclesfield where adders could be caught in profusion. One area was that of Alderley and Alderley. Woods.

The celebrated author and philosopher, Sir Kenelm Digby used to feed capons on only vipers flesh. These were kept entirely for his wife`s diet,she,Venetia Stanley, had a daily meal of three capons and was noted as one of the foremost beauties of her day; she was reputed to look 20 on her fiftieth birthday in 1660.

It is necessary to delve into the realms of myth and magic to know the adders primitive properties.
Men who had learned of their own mortality and had seen the snake `shed her enamelled skin…weed wide enough to wrap a fairy in" and emerge reborn,longer,stronger, larger than before. Even more beautiful than before. Birth – rebirth.Immortality and eternal youth.

West Africans eat the heart of the lion to acquire its bravery; the blood of the cheetah to steal its fleetness of foot and would under no circumstances eat the flesh of a turtle.

This principal is one which goes back to the dawn of history and recorded medicine. From India to Gaelic traditions is this same faith in the serpents power. Snakeskins boiled in wine for rheumatism in Cheshire. Tincture of Adder for the plague in London. Adder broth for the bite of mad dogs in Paris and I have seen in 1980 the skin of an adder dried and hung over the hearth to bring good luck and health to those in the house. One in the New Forest and another on the Isle of Skye.

Old wives tales? Ten thousand years of myth and magic? Modern methods of analysis have isolated the complex and dangerous proteins of adder venom and that of many other poisonous snakes. There are enormous possibilities that they could provide the answer to a number of previously incurable complaints and one American company is very involved with the production of vaccine and sera from snake venom. In strengths which approach almost homeopathic doses they have had success diseases for which there have previously been no cures.

I fear it will, under modern legislation, take a long time to reach the market. [26]

In 1972 *The Times* reported, in an article titled `Country Tales of snakes and adders` by Alison Ross:

If you want to avoid adders, avoid fires in likely adder places. The old adder-men (there seem to be none left now) knew this and used to light a small fire inside a ring of stones. They would wait near and pounce on adders as they came to investigate. The "adderers" would pick them up by the tail,bash the snakes over the head until they were dead and then sling them round their own necks. They killed the snakes for their fat". (27)

Thanks very much to Bob Skinner for significant help with this essay.

REFERENCES

1. Atwell. *The Philosophical Transactions of The Royal Society.* Some observations on a

man and a woman bit by vipers. (1734) p.324 ff

2.   W. Ellis. *A compleat system of experienced improvements...*(1749) pp. 105-7

3.   *Medical repository*. On the effects of oil in cases of bites of serpents. (1805) p.242

4.   J. Clare. Snakes – in John Clare *Selected Poetry and Prose* edited by M and R Williams (London: Methuen and Co. Ltd 1986) pp 142-145

5.   *Farmers Journal* 1834

6.   *Manchester Guardian* Local Notes and Queries October 12th 1874.

7.   T. Hardy *Return of the Native*. Project Gutenburg on-line edition. (2006)

8.   *The Life of Thomas Hardy* . Florence Hardy. (Studio Editions Ltd 1994) p. 18

9.   National Parks website   http://www.nationalparks.gov.uk/ch_nf_historical

10.   Southern Life website   http://www.southernlife.org.uk/brusher_mills.htm

11.   D. Stagg *Snake Catchers of the New Forest* (New Forest Association 1983) pp 3-11

12.   R.Muirhead. Some Strange Snake Stories. *CFZ Yearbook* 1998 p.186

13.   C. Kingsley. *On Bio-Geology* (1871)

14.   *Hampshire Advertiser* March 11th 1900 p.8

15.   *Portsmouth Evening News* May 5th 1904 p. 2

16.   *Exeter and Plymouth Gazette* October 14th 1905 p. 5

17.   *Country-side* May 25th 1907 p. 43

18.   *Country-side* October 20th 1907 p.351

19.   *Western Morning News* May 19th 1922 p.6

20.   E-mail from B. Skinner to R. Muirhead September 2012

21.   *The Field* November 27th 1943; p. 563 C. Millson, Tales of Old Hampshire,   (Newbury, 1980) , pp 78-79

22.   The Snake Catcher (Brusher Mills) http://www.southernlife.org.uk/brusher_mills.htm

23.   *Country-side* August 5th 1905

24.   D.Stagg *op cit* pp. 11-13

25.   Rev J.E. Kelsall. The Reptiles of Hampshire, And the Isle of Wight in *Papers and Proceedings of the Hampshire Field Club* vol 3 (1897)

26.   C.Hadfield . *Hadfield`s Macclesfield Memories* (1984) p. 47

27.   A.Ross Country tales of snakes and adders in *The Times* September 2nd 1972 p.14

The Amazing Animals of Pakistan

Raheel Mughal

There aren't that many superlatives in the English Dictionary to describe Pakistan's beautiful wildlife; it is simply amazing. Pakistan's flora and fauna constitute approximately 188 species of mammal, 666 species of bird (both resident and migratory), 174 species of reptile, 16 species of amphibian and 525 species of fish. When it comes down to plants, the land of the five rivers has 5,000 wild plants, of which approximately 372 species are only found in Pakistan.

However, wildlife tourism is - sadly - highly under-developed. Illegal hunting and poaching is frequent and there is very little - if any - funding to provide an infrastructure to enforce any wildlife protection. It is very frustrating to think that these fantastic animals are not looked after properly and they are suffering as a result of human greed in all its forms.

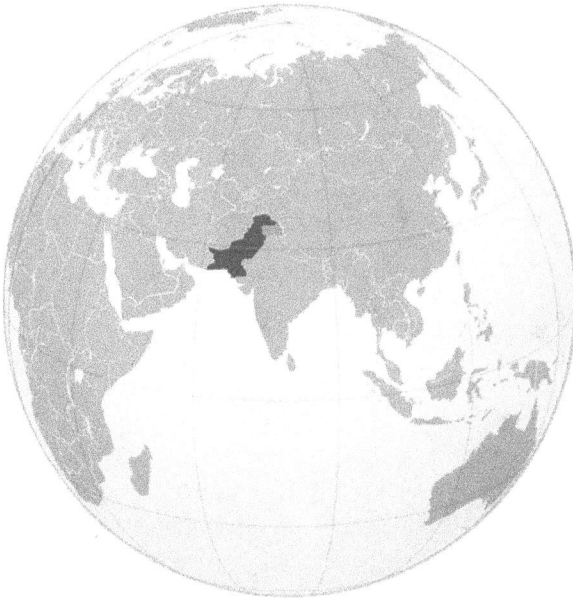

One of cryptozoology's main constituents is the study of rare and elusive animals. Pakistan has many, but as it is virtually impossible to include them all in this research report, and because this topic is largely beyond the scope of this work, I have decided to include a small selection of the most unusual and in some ways unique denizens of the four provinces of Pakistan (North West Frontier Province, Punjab, Balochistan and Sindh). Moreover, I have also written a research report on the mysterious yet beautiful snow leopard (also in this yearbook).

Asiatic Black Bear (*Ursus thibetanus*)

Also called the moon bear because of its distinctive yellowish-white chest patch, it is often found foraging in trees for nuts and fruits.

It is approximately 1.3 - 1.9m (4 ¼ - 6 ¼ ft) in length and weighs 100-200kg (220 – 440 lbs), respectively. All black, except for the white chest patch, the Asiatic black bear has tufted ears (giving it a rather untidy appearance) and a short tail. They have strong back legs allowing them to stand and walk upright unlike other bears found around the world, which can only rear up on their legs to survey their surroundings.

Asiatic black bears live solitary lives, with males and females only pairing up during courtship, after which they part ways. Female Asiatic black bears have a gestation period of 8 months, producing a litter of 1-3 cubs, which stay with their mother until they are old enough to fend for themselves.

Other than in Pakistan, the Asiatic black bear is also found in China, South East Asia, Eastern Russia and Japan. In some areas, such as in Pakistan, it is known to raid food crops and in the process kill humans. Sadly much of this bear's original habitat has been lost due to illegal felling of trees to make way for the leisure industry.

Moreover, the Asiatic black bear has also been persecuted in other ways. For example, it is the bear species used in bear baiting, the illegal practice where bears are forced to dance for the public, and if they don't they are beaten and tortured by their owners. These bears aren't fed and are forced to live in small claustrophobia-inducing enclosures where they are often poked

and prodded by their owners and their families. Also, in China, Asiatic black bears are often hunted for their body parts for use in oriental medicines.

As with all animal derived medicines there is no scientific proof that they do anything; it is just a waste of time resulting in the slaughter of an innocent creature just getting on with its own life. They are highly vulnerable to environmental damage, and they are now found in small pocket populations in forested hill regions.

Markhor (*Capra falconeri*)
Markhors are a species of tough mountain goat that inhabit the cold mountain ranges of

Central Asia, and in particular Pakistan where it is regarded as a sacred animal by tribal people of the North West Frontier Province. So much so, that the markhor has been given the status of the National Animal of Pakistan. Male markhors are characterised by their long corkscrew-shaped horns that can reach up to 1.6 m (5.25 ft) in length, and dominant males use these in duals during courtship rituals to retain control of the harem.

Markhors are approximately 1.4 - 1.8 m (4.75 – 6 ft) long and weigh approximately 32 – 110 kg (70 – 240 lbs). They have long shaggy manes with thick white and grey coats that are used primarily to keep the animals warm. Sadly, markhors are sought by trophy hunters and are also extensively killed for their meat and hides.

Shaheen (*Falco shaheen*)
The shaheen, the national bird of prey and also the motif for the Pakistan Airforce (see inset

below) , is essentially a sub-species of the larger, well-known peregrine falcon (*Falco peregrinus*) and is found largely in Pakistan, in addition to small populations living in India and Sri Lanka. A bit smaller in size than its cousin (the peregrine is 36-48cm), the shaheen also sports a ruby-red front making it a somewhat colourful bird of prey compared to the peregrine. With scimitar-shaped wings and a short tail, the shaheen is a living aerodynamic fighter jet, and silently swoops on to its prey with an amazing power dive. In most cases, prey items have no chance of escape. Shaheens make nests on cliff ledges as well as mobile phone masts.

Mobile phone masts are of great concern to some zoologists and biologists who believe that this may cause the nestlings to develop certain forms of cancer and genetic abnormalities at the embryonic stage, as well as later on in life. But like most theories this is not proven, although there is circumstantial evidence to suggest that this may happen to certain individuals. Therefore, more investigation into this topic is needed in order to help not only this beautiful bird of prey if these issues do cause a problem, but also other birds of prey that routinely nest on mobile phone masts.

Indian Pangolin (*Manis crassicaudata*)
Resembling a "walking pinecone", the Indian pangolin - or scaly anteater as it is also known -

is found mostly in the dry and arid scrub forests of Pakistan, with pocket populations living in India, Sri Lanka and some parts of South East Asia. However, recently they have become very rare in Pakistan.

The Indian pangolin is a ground dwelling pangolin species unlike its more arboreal relatives in Africa and parts of Asia. It feeds mainly on ants and termites, which it gathers with its extremely long and sticky tongue, swallowing them whole and grinding them up in its stomach. Nevertheless, they are also known to take in beetles, invertebrates and worms as part of their diet. The most prominent feature of the Indian pangolin (and with other pholidata) are their sharp-edged scales which are made of horn, and cover most of the outward facing parts of the body, head and tail. The scales offer a form of protection as well as camouflage. The scales are tilted at the base by muscles in the animal's skin and are replaced regularly.

Unusual for mammals, pangolins (other than having an outward appearance of small prehistoric reptiles), have some other characteristics which are not largely found in mammals. These include the ability to roll themselves into an armoured ball in times of danger, and the ability to run at short bursts of speed to evade capture, despite having layers of armour on at least 80% of their body.

Once, it was found right across the Indian subcontinent. Sadly, however, it is now found in a few small locales mostly living on the fringes in brush country or lowland forest habitats. It largely avoids encounters with mankind, which views the Indian Pangolin as either a tasty delicacy or as a medical ingredient to cure all manner of ailments from migraines to stomach ulcers, so much so that these activities have contributed to the animal's decreased numbers.

Desert Hedgehog (*Paraechinus micropus*)

The desert hedgehog is a solitary cousin of the European hedgehog, and comes in different colours. Some have banded spines of dark brown, with either black, or white, or with yellow.

As their name suggests, the desert hedgehog is found in dry desert regions in Pakistan where it builds deep burrows to escape the relentless heat of the day, which can sometimes exceed 35° C. During this time they lie in burrows 1-2m (3-6ft) deep, which they dig out themselves. They come out at night, when they prey on frogs and invertebrates. These animals are the nearly the same size as their European counterparts, only being slightly smaller at 14-27cm (5.6 – 13.5 inches) and weigh approximately 435 g (1Ib). Due to large-scale civil engineering projects (dams, reservoirs and highway construction) it is feared that the population of this animal may be affected in the future.

Gharial (*Gavialis gangeticus*)
The gharial is a large river crocodile, the male of which sports a pot-like cartilaginous structure on the tip of its nose (called a "ghara" in Urdu, essentially meaning "pot"). This is used as a resonator for buzzing sounds, which are used during courting. The male gharial is 6m (18ft) long, with females being 4m (12 ft), and this crocodilian is normally an olive-brown colour. It has webbed feet and a laterally flattened tail making it an excellent swimmer.

Its diet consists largely of fish, which this crocodile is superbly adapted to hunt with its long snout and razor-sharp teeth, and it makes quick work of any slippery river fish that happen to be in the vicinity of this lake monster. Nevertheless, the gharial's diet also consists of the odd bird or amphibian, and juveniles feed on a variety of insects, frogs and crustaceans among other things.

In Pakistan, the gharial is considered an endangered species by WWF Pakistan, and other environment agencies. It is thought that pollution and dam building have largely contributed to the large-scale demise of this fascinating and once widespread crocodilian. Gharials are also known from other countries of the subcontinent such as India and Bangladesh.

Indus River Dolphin (*Platanista gangetica*)

An extremely rare and critically endangered freshwater dolphin, this species is also one of the most elusive on Earth. Known locally as the susu or Sindh dolphin, or more commonly as the Indus blind river dolphin, this is the only known cetacean without a crystalline eye lens, making it effectively blind. Instead the dolphin uses echolocation to locate its prey (mainly fish and crustaceans), which it hunts in the silt laden water of the Indus.

The Indus river dolphin is described as being approximately 2.1-2.5m (7-8 ¼ ft) in length and weighs 85 kg (185lbs). It has a grey colouration at the top, and is pink below, with a long, thin gharial-like beak containing a range of protruding front teeth, which – incidentally - like the gharial are used as an interlocking cage to hold fish so that they don't swim away. The dolphin also possesses a triangular ridge on its back and also has broad, paddle-shaped, rugged flippers and large, wide tail flukes.

The Indus river dolphin has a flexible neck, which it can turn at right angles so that it can scan the area for prey with echo-locating pulses.

The Indus river dolphin lives in small groups of 4-6 individuals, though as many as 30 individuals have been recorded by WWF Pakistan. They feed in the Guddu and Sukkur barrages of the Sindh province. There are fears that there are only 1,000 individuals left within a 1200 km stretch of the Indus river system. It is thought that pollution, the building of the Guddu and Sukkur dams, and the barrages have all contributed to restricting the movement of these highly sociable animals. Moreover, the illegal slaughter of individuals for parts or food is rampant in the poorer areas of Sindh. The Indus river dolphin has a close relative in India, known as the Ganges river dolphin.

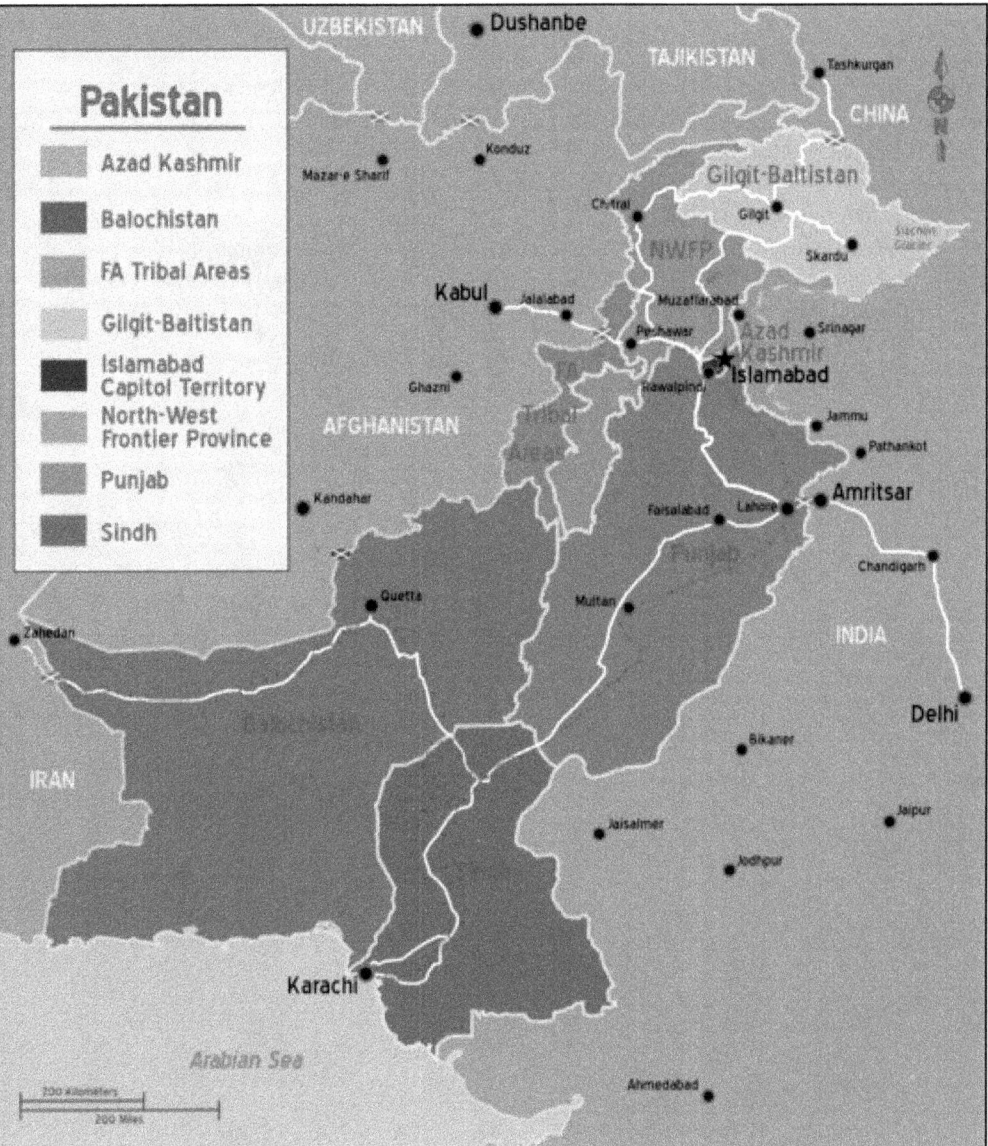

Pakistan

- Azad Kashmir
- Balochistan
- FA Tribal Areas
- Gilgit-Baltistan
- Islamabad Capitol Territory
- North-West Frontier Province
- Punjab
- Sindh

References

- Pakistan Natural History Museum Islamabad (Visited April 2011).
- Jackson. T (2010) *The World Encyclopaedia of Animals*, Anness Publishing Ltd, Hermes House UK
- Lonely Planet Guide (2004), *Pakistan & the Karakoram Highway*, 6th ed, Lonely Planet Publications, UK.
- Tritsch, M. F (2001) *Traveller's Guide - Wildlife of India*, Collins UK.
- Clutton-Brock, J (1995) *Mammals of the World,* Dorling Kindersley Handbooks, Dorling Kindersley UK.
- www.wildlifeofpakistan.org
- www.wwfpakistan.org

The Mpisimbi
– AN UNDISCOVERED BUT NOW-EXTINCT KING CHEETAH STRAIN IN EAST AFRICA?

Many mysterious African animals once thought to be legendary or wholly imaginary monsters solely confined to the realms of native folklore and superstition have ultimately been found to be real (albeit elusive) creatures that have successfully eluded formal scientific recognition until modern times. The mountain gorilla, okapi, giant forest hog, and pygmy hippopotamus were all dismissed as myths by Western science until the 20th Century. So too was another bizarre beast - the leopard-hyaena or nsui-fisi...until 1926.

UNMASKING THE NSUI-FISI
This was when a short letter penned by Major A.C. Cooper from Salisbury (now Harare) in Rhodesia (now Zimbabwe) was published by *The Field*, alongside his photograph of an extraordinary cat skin. Cooper believed the skin to be from a crossbreed of leopard and cheetah, which had been trapped at Macheke, about 62 miles southeast of Salisbury, but he was astonished by the exceptionally ornate markings adorning its golden-yellow coat, which were unlike those of any cat previously recorded by science. Upon its flanks and upper limbs they consisted of graceful curved stripes and abstract blotches, whereas a series of longitudinal stripes ran from its neck and shoulders along the entire length of its back to the upper portion of its tail, and a succession of thick black stripes encircled the remainder of its tail. Also of note were its non-retractile claws (a cheetah characteristic), and a mane-like ruff round its neck.

This wonderful creature matched traditional tribal descriptions of a strange monster termed the nsui-fisi ('leopard-hyaena'), which was fervently claimed by Rhodesian natives to be the rare, exotic progeny of liaisons between leopards and hyaenas, as it was said by them to be as lithe, swift, and cunning as a leopard but boldly barred like the striped hyaena *Hyaena hyaena*. Such an identity was zoologically impossible, of course, but the fact remained that Cooper's mysterious big cat was still unexplained.

Not surprisingly, it soon attracted the attention of felid specialist Reginald Pocock, from the British Museum (Natural History), who identified it from Cooper's photograph as an aberrant leopard. (Interestingly, during the early years of the 20th Century, cats just like it had apparently been well known to locals in the Mazoe area of Rhodesia's northern region, where they were referred to as Mazoe leopards.) Once he was able to examine the skin itself, however, he swiftly changed his mind, recognising it to be from a cheetah - albeit one with dramatically different coat markings from the familiar polka-dot pelage of the normal cheetah *Acinonyx jubatus*. In 1927, Pocock announced that the skin represented a new species, which, due to its regal appearance and vaguely leonine mane, he dubbed *Acinonyx rex* - the king cheetah.

Painting of the king cheetah type specimen as imagined in life, from *PZSL*, 1927

During the next few years, several other king cheetah skins were obtained - all from a triangle of terrain enclosing eastern and southern Zimbabwe, northern South Africa, and eastern Botswana - but as the number of skins increased, it became evident that some of these were intermediate in appearance between normal cheetahs and the first king cheetah, documented by Major Cooper. In other words, there was no longer a clear morphological demarcation line between spotted cheetahs and the striped king cheetah. This could mean only one thing - the king cheetah was not a distinct species in its own right after all. Instead, it was merely a freak mutant variety of the normal cheetah - which Pocock conceded in 1939.

As a result, interest in the king cheetah waned, and after a time reports of such specimens rarely emerged. Indeed, by the 1970s some zoologists had begun to fear that this handsome striped strain had died out, but during the 1970s a king was filmed living with normal cheetahs in the Kruger National Park.

Today, there is no doubt that the king cheetah is indeed alive and well, with several specimens having been born within litters of normal spotted cheetahs in captivity in South Africa.

THE GENETICS OF A KING

Since its scientific discovery, there has been a great deal of conflict concerning the king cheetah's taxonomic status. At first it was believed to be a hybrid of leopard and cheetah, then a valid species, and ultimately a mutant form of the normal cheetah. In May 1981, however, the de Wildt Cheetah Breeding and Research Centre of Pretoria's National Zoological Gardens was able to examine this issue in a thorough manner, when a king cheetah was unexpectedly born to a pair of normal cheetahs here. A few days later, moreover, a second king was born, this time to the sister of the first king's mother.

Two normal spotted cheetahs

These fortuitous events duly initiated a programme of monitored breeding conducted at the centre, in order to determine the genetic basis of the king cheetah phenotype. By 1986, it had become clear that a recessive mutant allele was responsible, equivalent to the recessive allele producing the blotched tabby coat pattern in domestic cats. In other words, only cheetahs with two copies of the 'king' allele are kings. Cheetahs with one copy of the king allele and one of the normal (wild-type) spotted allele, or cheetahs with two copies of the spotted allele and none of the king allele, are normal spotted cheetahs.

Moreover, in a joint paper published by *Science* in September 2012, a team of American genetics researchers from several different institutes revealed that they had identified the specific gene responsible for the king cheetah's striped coat pattern and the blotched coat pattern in domestic tabby cats. Both are caused by a recessive mutation in a gene dubbed Taqpep by the researchers.

And so, one mystery concerning the king cheetah is a mystery no more - but there are others that still await a solution, and none is more fascinating than the following example.

UNMASKING THE MPISIMBI?
What makes the king cheetah so memorable in addition to its incredibly beautiful coat is its extremely limited distribution. Many freak mutations of coat colour or patterning in mammals are spontaneous, i.e. they can arise abruptly in any population of a given species, regardless of geographical location. Yet whereas the typical spotted cheetah occurs in southern, eastern, central, northern, and western Africa, king cheetahs have never been reported conclusively outside southern Africa – or have they?

There are two possible and potentially extremely significant exceptions to this widely-assumed rule (and three, if we consider some remarkable evidence that I have lately uncovered for the erstwhile presence of at least one king cheetah specimen in the wild not anywhere in Africa, but instead in Asia; however, that, as they say, is another story, whose telling must wait for another time!).

The first possible exception to the king cheetah's strict zoogeographical limitation to southern Africa is a king cheetah skin that in 1988 turned up in the West African country of Burkina Faso. It supposedly came from a specimen that had been shot by a poacher in the northern end of the Singou Total Fauna Reserve. Some researchers wonder whether this mystifying skin is one that in reality originated in southern Africa but which later travelled northwest via itinerant poachers or other skin traders. Alternatively, however, could a 'king' strain of cheetah have spontaneously arisen in West Africa?

As for the second putative exception to the rule of the king cheetah being exclusively southern African in distribution, this is one that has never been publicly revealed – until now. It features an extremely obscure East/Central African mystery beast known as the mpisimbi

In 1927, *Chambers's Journal* published a fascinating article on East and Central African mystery beasts entitled 'On the Trail of the Brontosaurus and Co.'. It was written by 'Fulahn', the pen-name of Captain William Hichens - a man whose name should already be familiar to mystery cat aficionados. For he was none other than the Native Magistrate at Lindi, Tanzania, during the 1920s and 1930s who investigated a succession of particularly gruesome murders there attributed by the local people to a giant brindled mystery cat known as the nunda or mngwa.

King cheetah head (above) and normal spotted cheetah head (below)

Most of the cryptids documented by Hichens in his *Chambers's Journal* article are relatively famous ones, with one notable exception. Contained within his account are a couple of tantalising lines that have fascinated and frustrated me in equal measures for many years:

> But such are the mystery animals. There are others – the mpisimbi, the leopard-hyaena, which eats sugar-cane, and which I have hunted many a weary night without success;

Despite numerous searches, I have never been able to uncover any additional information concerning this enigmatic creature. So what exactly *is* the mpisimbi?

King cheetah (Alan Pringle)

The above-quoted lines offer no morphological description whatsoever of the mpisimbi. Conversely, its name's English translation – 'leopard-hyaena' - is very intriguing, because it corresponds precisely with the English translation of the king cheetah's native South African name, nsui-fisi. Moreover, Hichens's unusual claim that the mpisimbi eats sugar-cane adds further to a putative link between the mpisimbi and the king cheetah, because in a second article, published under his own name a year later in *Wide World Magazine*, Hichens stated: "The Nsuifisi, or striped cheetah...was also reputed to be a raider of grain and sugar-cane".

Of course, as Hichens went on to discuss, because cheetahs are carnivores it seems improbable that they would raid grain-plots. And even though hyaenas are notorious scavengers with an extremely catholic diet, they are not known to attack standing crops, but they will certainly devour cooked grain, vegetables, and even boiled flour.

Such considerations and qualifications, however, are not significant with regard to the cryptozoological mystery under review here. What *is* significant is that both the mpisingi and the nsui-fisi were claimed, rightly or wrongly, by the native tribes in their respective, separate areas of Africa to consume the very same unexpected foodstuff – sugar-cane.

Is it conceivable, therefore, that the mpisimbi and the nsui-fisi are indeed one and the same creature – namely, the king cheetah? If so, it suggests that at some time in the distant past, striped cheetahs did exist in East and/or Central Africa – although, with no modern-day reports of such beasts on file, even if they once did exist there they seemingly no longer do. Put another way: whatever it may have been, tragically the mpisimbi is now apparently extinct.

Of course, sceptics may well claim that this is all supposition, but the presence of those brief lines regarding the mpisimbi in Hichens's article means that the possibility of mpisimbi and king cheetah synonymity, however remote it may seem, cannot be discounted.

Moreover, who can say whether, in the future, a king cheetah or two will not spontaneously arise in the East African population of the normal spotted cheetah? That is, after all, what spontaneous mutations do!

At present, however, the mystery of the mpisimbi's zoological identity remains yet another enigma in the eventful history of Africa's extraordinary striped cheetah - the once (and future?) king.

This article is an adapted excerpt from Dr Karl Shuker's newly-published book *Cats of Magic, Mythology, and Mystery: A Feline Phantasmagoria* (CFZ Press: Bideford, 2012).

Lion hunt is a shaggy dog story

A LION hunt was called off last night when a zoo expert revealed that mystery paw prints were made by a large dog.

Police called in marksmen and a helicopter to help search a wood after an officer in a patrol car and a housewife claimed they had seen a lion.

Mrs Gwen Shipman, 49, of Cuffley, Herts, said: "I saw it in the grass next door. There was no mistaking it."

Police kept two officers in the area late last night "just in case."

The 'Lion' Of Essex And Other British Big Cat Scares

Neil Arnold

The mane attraction/distraction (delete where applicable!)

What's tan coloured with a sweeping mane, has got a pristine set of gnashers, grooms itself daily, appears in the newspapers for no reason and is often seen basking in the Essex sun? "One of those z-list celebs from *The Only Way Is Essex*," I hear you cry. No silly, it's the alleged lion of Essex – a formidable prowling pussycat native to Africa and yet said to have stalked the wilds of rural Clacton during late August 2012.

Sadly, the only way for a lion wasn't Essex, and the maned monster didn't exist after all. What a surprise! And, like so many other 'big cat' scares that have been known to attract the attention of the local authorities, this particular farce was a sum of many parts – misidentification, hysteria, newspaper hype, hoax and above all, repetition. So many similar cat flaps had gone before it, spawning laughable headline after laughable headline, bringing the camo' daubed researchers out from their holes, and the journalists from behind their desks on a day when real news was slow. The police – with the health and safety of the local public in mind, spent a lot of time and money buzzing the skies with helicopters, and scanning the ground with surveillance equipment but to no avail, and after a day the search was called off. It all started, or so they say, when Rich Baker, 39, was walking with his two children when they saw a man who was running and screaming, "It's a lion!"

Mr Baker told *The Daily Mail*,

> "The lion was in the middle of the field with its back towards me. When someone else screamed it turned around and you could see it from the side. I grabbed my children and we ran towards our caravan. It was one million per cent a lion. It was a tan colour with a big mane, it was fully grown..."

The frightened witness also told reporters that at least a dozen or so people had also observed the cat and at 7:00 pm on the evening of Sunday 26th August it was claimed that the lion had been photographed, but all the photographs that had been circulated on the internet were nothing more than a hoax until Jill and Steve Atkin from Lincolnshire came forward to say they'd spotted the beast lazing in a field and so snapped the creature. Sadly, the photograph turned out to be a domestic cat called Teddy Bear, far removed from the king of the jungle said to be prowling St Osyth on the Bank Holiday weekend. Even so, the photograph made several headlines but was simply proof as to how eye witness observation can be extremely unreliable.

I stood back from the crowd and watched in amazement as the television reporters gathered in Clacton like excited children, and they prayed for a lion to turn up. One lady reported how she'd seen a fawn-coloured cat with a white chest and a long thick tail – certainly not the description of a lion.

There are no lions, or, for the record, tigers or cheetahs or jaguars roaming the wilds of Britain. I've investigated reports and collected evidence for more than twenty-five years – it is now a full-time job - with regard to the existence of smaller cats in the UK woods, and believe me, this island of ours is very much home to a variety of species of unusual cats, but none of these are bigger than a leopard. Of course, the existence of 'big cats' in our midst has long been debated and so it may seem a rather sweeping statement to dismiss lions and yet believe in leopards. Either way, such creatures have most certainly cemented themselves into British folklore alongside those half-hinted ghosts, blurry UFOs and occasional 'monsters' such as Nessie and the like. However, the evidence, which many sceptics seem to overlook in their quest for ultimate dismissal of such animals, is in abundance and yet only recently gathered for analysis. The big problem with UK 'big cat' stories is that headlines such as *The Sun*'s 'The only prey is Essex,' or the *Daily Mirror*'s 'Dear Kitty, Kitty' simply make a mockery of all the hard work being done by researchers who know for a fact that large animals such as black leopard, puma and the once native lynx, inhabit Britain.

The Essex lion story didn't make sense, but my 'phone began to ring continuously, the emails popped into my inbox at a regular rate, and friends of mine began texting me to ask me what I thought about the lion on the loose. I was bewildered as to how so many people could take the story so seriously and also confused as to how many sceptics began to mock it and in turn connected it to the sightings of other large cats in the UK. As soon as the Essex story died a death, the blogs began to appear, one journalist eager to dismiss such reports of 'phantom cats' as nothing more than hysteria akin to India's Monkey Man and Victorian London's Spring-Heeled Jack. Of course, what a majority of the public and the press didn't realise was that six days before the lion story took off a chap named Jack Stone had emailed me to say that during the middle of August, during dusk one evening, he'd been driving with his girlfriend near St Osyth when a large, fawn-coloured cat had crossed the road some thirty metres in front of the vehicle. Immediately Jack's girlfriend gasped, "It's a lioness," but Jack, unlike so many witnesses involved in the Essex lion debacle realised what he was seeing.

> "A puma crossed the road – it was around four-feet in length, very low to the ground, and moved with grace but speed, and the tail is what stuck in my mind – it was long, thick and curved upwards at the end."

Jack had spent time with family in the United States several years ago, and during one jaunt through the Californian wilds had been fortunate enough to observe a mountain lion one night.

The animal he saw cross the road in St Osyth looked exactly the same.

It's possible that the Essex lion scare may have initially sprung from an encounter or two with a puma – this wouldn't be the first time this had happened.

The Surrey lioness and others
In the late 1950s and early '60s the 'Surrey puma' panic began with several reports of an alleged lioness on the loose. The press had a field day with the stories at the time, the Surrey puma, alongside the beasts of Exmoor and Bodmin, becoming the king mystery cats of their field, but at the time so many people, lacking knowledge of exotic cat species, were keen to speak of a large cat with a "sandy coloured coat," which to many untrained reporters, and even police officers sounded like a lioness. The authorities took to the fields of Shooter's Hill in the '60s but never flushed out the animal. One police officer commented that the animal may well have been a cheetah because even on his police motorbike he couldn't keep up with the elusive cat! The Surrey puma caused great debate, although not many realise that one of the first reports of a large cat roaming Surrey comes from the 1700s from the area of Coulsdon. Authorities blamed a "wild dog" for the slaughter of livestock even though all the sheep had been killed in a certain manner – a bite to the throat and then the rasped fleece and bones.

Who would have thought that more than half a century after the Surrey puma flap police would still be combing the woodlands and fields of England in search of a similar entity? Every year I receive several reports from Essex of large exotic cats, mainly the already mentioned puma, and also black leopard. Epping Forest, Ongar Marshes, Witham, Brentwood, and Chelmsford just five of many a location where 'big cats' have been sighted. Although the photographs of the Essex lion were a non-event there is a possibility that a puma had been present but by the time the police had arrived it had slunk away into the shadows.

A male lion isn't an elusive animal – in the wild they seek a pride, are large enough to stand out like a sore thumb, and would seek large, possibly human prey if out of their natural environment. Writing for *Parascience* magazine in the summer of 1999, researcher Jonathan Downes commented, 'the earth-shaking roar of a lion can be heard up to five miles away…'

Rumours persisted that the Clacton critter had escaped from a circus that had been stationed in the area a few weeks previous, but these whispers were unfounded but also typical of how UK 'big cat' folklore works. As soon as someone sees something out of place they begin to blame a circus, or local zoo, and in most cases of lions escaping from zoo enclosures or private collections, they are either recaptured or shot dead. The fact the search for the Essex lion was called off was proof enough that no massive animal had been seen, the police were merely covering their backsides, although the Hampshire 'white tiger' scare from 2011 left me with little faith in the authorities. Once again the police had caused chaos, and in a sense sparked a witch hunt by evacuating a local golf course, ordering the public to stay indoors, and buzzing the fields close to the M27 when witnesses claimed they'd spotted a white tiger, which turned out to be nothing more than a cuddly toy.

Again, there appeared to be no common sense in the eye of the storm. Firstly, how on earth would a white tiger end up in Hedge End? Secondly, why didn't the animal stir as the authorities closed in? And thirdly, wouldn't such a wild beast be visible through binocular lens? Only when the thermal imaging cameras failed to register any heat emitting from the 'animal' did the officers become suspicious. Once again, the press had a field day, and once again the sceptics scoffed at the possibility of large cats roaming the UK. Even so, in the November of the same year another lion scare took place, this time in West Yorkshire after a witness claimed to have seen a large cat near Shepley Station. A two-hour search involving twelve officers and a helicopter drew a blank.

Reverting back to Essex, the county has had lion scares before. In 1996 there were reports of a male lion on the loose around Maldon, but again, rarely, if at all, was a mane described, but in most of the sightings witnesses mentioned an animal that was "sandy coloured" with a long, thick tail. A police officer visited several of the locations where the 'lion' had been seen, but was once again quick to dismiss the reports as nothing more than a large domestic cat.

Panic on the streets of London...
In 1994 police were called out to investigate the 'lioness' of Winchmore Hill in north London. Witnesses 'phoned the local authorities to speak of their encounter with a large, sandy-coloured cat that had been seen walking along a canal. A helicopter whirred over Palmer's Green and marksmen from London Zoo took to the undergrowth armed with tranquiliser guns but once again, the beast failed to rear its head until the *Daily Telegraph* of March 12th commented that the loose 'lion' had in fact been Bilbo, a large tomcat belonging to a Carmel Jarvis. A photograph snapped by another witness – who was adamant he'd seen a wild animal – proved once and for all that the lioness had been nothing more than a moggy. Even so, Robert Delane claimed he'd seen a mountain lion not far from a stretch of the London Underground in the June of '94, and despite London Zoo's Doug Richardson commenting that it was very likely that a puma could survive in the area, the reports died a death once Bilbo had been named as the culprit. Mind you, in 1943 a lion was most certainly on the loose for a brief time in London. The *Sydney Morning Herald* of October 22nd 1943 reported that a large cat had been seen by a woman in the vicinity of Clapham Junction. Officers were alerted to the animal, which had escaped from a box-car, by the shrieks of the terrified witness. The cat had escaped its confines when a train pulled in to the station and the animal had leapt on to the platform. It then casually strolled up to the nearby office window and glared in at the inspector who almost had a heart attack on the spot. The cat then ambled along the railway track. Southern Railway Home Guard officers were called in to action accompanied by police, and watched by hundreds of unnerved spectators, they ushered the cat into a railway pit and awaited the arrival of two lion keepers who recaptured the cat. Three years previous residents of Cuckoo Hill Road in Middlesex were in uproar (excuse the pun!) after George Thomson had been keeping a lion named Rona in his back garden. In 1968 a four-year old lioness named Beatrix escaped from its enclosure at Chessington Zoo. As her keeper was tending to the four cubs Beatrix slipped by him but was eventually recaptured after being tranquilised. Weirder still, in the same year a leopard cat escaped from the same zoo and wandered into a flat in Richmond!

In 1976 a Surrey woman named Poppy Hull was accosted by a pet lioness named Shane

(strange name for a female!) which had leapt from its enclosure – the back garden of its owner, a taxi driver named Ronald Voice – due to its disliking for the woman's leopard-skin coat! The 14 stone cat did not hurt the woman, but left her visibly shaken. *The Times* of 24th March '76 reported that Mr Voice was not in court on the 23rd but 'the judge granted an injunction to Woking Borough Council' banning Mr Voice from taking his pet lion into "any street or public place unless she is securely caged and chained and attended by himself or a keeper."

1976 was the year when the Dangerous Wild Animals Act was passed by the government (this bill meant that any owners of such cats and other exotic animals would have to pay for a license to house their 'pets.' A number of people released their animals instead, and this is the main reason as to why cats such as puma and black leopard roam the UK wilds today) but it didn't stop a London man from purchasing a lion to guard his house whilst he was away on holiday. One afternoon a man collecting for a charity was knocked down by the animal. The frightened, but unharmed man ran to the local police station and the cat was sent off to London Zoo. In 1987 *The Sun* reported on the 'Park Hunt For A Lion' after there had been a handful of reports of a 'big cat' in the Staines area of Middlesex. According to the report of 26th June, police were taking the sightings very seriously after 'fur found in Laleham Park was identified as genuine lion's fur.'

Researcher Marcus Matthews who wrote the fantastic book *Big Cats Loose In Britain*, investigated the case and spoke to the Chief Inspector of the Metropolitan Police at Staines who told him that the fur was probably from a dog. Mind you, in 1988 a tiger was said to have been seen in London's Edgware district by David Corbel who reported that the animal "had a black body, ginger hindquarters and white paws," although these details fit no known cat species, but the fact the animal was perched up in a neighbour's tree suggested to some that the animal had in fact been a melanistic leopard, but a police search found nothing. In 1927 a large cat escaped from its cage at Camden Town. The leopard slipped out of its Park Street enclosure during the early hours but was finally recaptured the following afternoon by its owner when the animal was seen sitting in a garden in the company of a small child. The child's mother raised the alarm and contacted owner George Palmer who said that the animal was not a threat. In December 1899 police were called to investigate the escape of a black leopard in the vicinity of Woolwich. The incident, reported on by *The Times* of Dec' 8th came about when the cat had escaped from its enclosure aboard a ship docked at the Royal Albert Docks. The cat was only recaptured when the crew of the vessel lit newspapers and threw them into the building where the animal was loitering. The cat rushed back into its cage.

The last time police were called out to investigate an alleged 'big cat' in London was in 2005 when Mr Holder of Sydenham claimed to have been attacked by a black leopard one night. This story appears to be nothing more than a hoax, and when I investigated the incident was more alarmed by the police presence – armed with Taser guns – and hysterical press – than any savage leopard. Of course, the fact that a lynx did turn up in a back garden in London's Cricklewood region in 2001, was proof that in some cases of so called beastly headlines, facts are stranger than the fiction, but as stated earlier, smaller cats, such as lynx, and also jungle cat and leopard cat, are still kept illegally in private collections. In 2011 a juvenile Amur leopard cat turned up in a London garden and was handed into Battersea Dog's Home and then eventually ended up at Heathrow Animal Reception Centre before being moved on to a zoo in Galloway. However, are stories of lions in our back yards completely unfounded? Well, surprisingly, no.

More beasts on the loose
In 1939 a hunt ensued to find a leopard that had escaped from its enclosure at Primley Zoo in Paignton, Devon. On January 10th 68-year old Primley Zoo keeper Jack Hawkins was severely mauled by the cat which escaped into the night. The area was cordoned off by more than forty police officers whilst the streets of the town were patrolled by civilian volunteers. *The Times* of 11th Jan' reported that 'messages were flashed on cinema screens warning the public, who were also advised to keep their children off the roads.' Farmers were also told by authorities to 'watch their flocks' and 'bus conductors in the area were instructed to warn all passengers dismounting at lonely spots.'

On January 12th the escaped leopard was, sadly, shot dead. According to the *Glasgow Herald* of January 13th 'It was killed by Major S.A. Yorke, of Starcross, officer commanding the 152nd (Devon) Light Anti-Aircraft Battery Territorial Army.'

The cat had apparently been held responsible for the deaths of six sheep during its hours of freedom. The animal was tracked by its spoor and found sitting under a bush. Major Yorke, who got to within nine-feet of the felid, shot it in the head and then to make sure fired another bullet into the shoulder of the cat. The newspaper report concluded that the animal was a seven-year old jungle-bred specimen that measured seven-feet in length and weighed 200lbs.

The most intriguing report I have come across regarding a lion on the loose in Britain comes from the pages of *The Sydney Mail* from November 16th 1889 which appeared under the heading, 'A lion hunt in a sewer.' The incident took place in Aston, in Birmingham on 27th September. The animal, not simply a domestic cat this time, had escaped from Wombwells' menagerie after a keeper had entered the cage to clean it out. The cat, 'a four year old black-maned Nubian lion' apparently slipped from its cage unnoticed until it reached the streets where 'thousands' of people at the local fair were said to have fled in terror. The creature bounded away and was probably as frightened as the public, and it made its way into a sewer after being pursued by two keepers. The lion had squeezed itself into the 2ft 6 inch diameter hole and although one of the keepers fired a revolver into the darkness – which was met by a ground shuddering roar - the animal refused to budge to the entrance where a cage awaited. The newspaper however commented that 'To allay the excitement it was reported that the animal had been recaptured, and crowds flocked to the menagerie to see a lion which was exhibited as the runaway.'

On the following Saturday however the news spread that the escaped lion was still on the loose, and two brave men employed at the menagerie decided that, armed with guns, they would enter the sewer in an attempt to force the lion into the cage. Although the men eventually ran out of ammunition they managed to persuade the cat to walk into a noose held by menagerie proprietor Mr Bostock. He roped the animal around its loins and somehow hauled it out of the drain, with the help of several policemen but the animal proved to be too big for the cage, its head sticking out allowing its ferocious jaws to gnash at anyone who put their limbs too close. By 4:00 pm on the Sunday the cat, now exhausted, was found a larger cage and despite being shackled, many members of the public were reluctant to descend the lampposts which they climbed to avoid the ruckus.

On the 3rd September 2012 the *Birmingham Mail* ran an article stating that over the years there had been many 'big cat' scares within the Midlands, particularly the dense countryside between Birmingham and the Black Country with Cannock Chase being a hot-spot area. The newspaper mentioned how in the 1970s businessman Lew Foley kept a lion in his back garden, although this didn't explain the sighting of a strange striped cat in the August of 2012 – it turned out to be someone dressed up as a tiger heading off to a fancy dress party! Even so, one story which was taken seriously regarding a lion on the loose was the case from Rossendale – fourteen miles northeast of Manchester - in the 1980s which somehow made the headlines when a teenager named Owen Jepson claimed to have seen just the back legs of a large cat-like animal that had leapt into the undergrowth. Hardly a reliable sighting and certainly nothing to suggest a lion although the county of Staffordshire has had its fair share of lion panics. The most known of these concerned the village of Hopwas, where in 1997 a family reported seeing a male lion during broad daylight. The police were quick on the scene but once again could find no trace of the beast despite the rumours that several officers had heard a tremendous roar. Although the roar was never explained, the legend dissipated rapidly when an owner of a large Old English Mastiff dog came forward to suggest that it had in fact been her dog – in the vicinity at the time of the 'lion' encounter – responsible.

One of the most known cases of a lioness escaping from its enclosure in England took place on October 20th 1816. A mail coach travelling through the Wiltshire countryside was attacked

by a large cat that had escaped from a menagerie – the beast was said to have clamped its jaws around the neck of one of the horses which was pulling the Exeter coach, and locals fled in terror. The attack was said to have ceased when the menagerie owner, and his brave dog (which was killed by the lion), distracted the lioness, which was eventually recaptured. The *Salisbury and Winchester Journal* at the time commented that,

> 'Her (the lion) owner and his assistant's followed her upon their hands and knees, with lighted candles, and having placed a sack on the ground near her, they made her lie down upon it; they then tied her four legs and passed a cord round her mouth, which they secured; in this state they drew her out from under the granary, upon the sack, and then she was lifted and carried by six men into her den in the caravan.'

This incident has become the stuff of folklore, and artist James Pollard depicted the scene in his painting, *The Lioness Attacking The Horse Of The Exeter Mail Coach*. In 1998 BBC News reported 'UK hunters try to bag Dartmoor lion,' after it was rumoured that a male lion had been seen loitering near the village of Wrangaton. The fuss began when a motorist reported seeing a cat-like creature which loped across the road in front of his vehicle before slipping into undergrowth. Armed police, accompanied by tracker dogs took to the woods and fields. The only trace of a large cat being on the loose came in the form of a paw print which had been cast by Robin Godbeer, a big cat specialist at Dartmoor Wildlife Park. In the same year several newspapers were quick to speak of the 'tiger' on the loose in South Wales. Police were called out to Garnswllt after three boys claimed that they'd seen a 'big cat' in the woods. Guess what? The police found no tiger. Four years previous it was reported that armed police were on the hunt for a lion near Basingstoke, Hampshire after several reports of a 'lioness' made by witnesses. A plane, using thermal imaging, was employed but no trace of the cat was found despite being sighted 'eight times in five days' according to an unnamed newspaper. During the same year a lioness hunt was carried out in the Yorkshire Wolds. According to the *Yorkshire Post* of 18th August 1994 'armed police launched a safari-style operation' after numerous sightings of a cat. One witness, Mrs Hutchinson, reported, "There is absolutely no doubt in my mind that what we saw was a lion," even though all other reports described a light-coloured cat with a long thick tail and with the build of a lioness – in other words a possible puma. Even so, a helicopter was drafted in from RAF Leconfield to aid the armed police who had sealed off two square miles of farmland within the hamlet of Ruston Parva but after a six hour search the lion/lioness failed to rear its head. Shortly afterwards it was rumoured that the elusive cat had left a spray of urine which, according to the Post, had been sent off for examination by forensic scientists. Matthew Brash, a veterinary surgeon from Malton, commented, "There was a definite wet patch where they said they saw the animal urinating…"

The tests proved inconclusive but as the Post commented on the 20th August, 'It's about time that Yorkshire had its own big cat legend.'

1994 was most certainly the year of the 'big cat' however; the year ended with a cat flap in Hertfordshire when a snow leopard had been seen after escaping its cage on a private reserve at Welwyn. Armed police (again!) took to the snow covered fields with their dogs in the hunt

for the cat, believed to have been a 16 month old cub named Tara. On this occasion the rumours turned out to be true and the cat was found sitting six-feet up a tree. She was darted and after becoming drowsy safely removed from the woods in a net.

In 2008 the city of Belfast was on high alert after a 'large, sandy-coloured' animal, believed to be a lion, was rumoured to be on the loose in the Cavehill Park area. The cat had been seen close to Belfast Zoo leaving many people pointing the finger at the park, but zoo keepers after a thorough inspection stated the animal had not come from their enclosure. Just a few days later a police spokesman told the press that there was no lion on the loose and that the animal seen by witnesses had probably been a large dog. This case echoes the hilarious story of the creature dubbed 'The Beast of Eczema' that made the headlines in 1999 after several sightings of a lion-like creature in the Yorkshire town of Barnsley. Police and animal welfare officers conducted a hunt for the 'lion' only to discover that the animal responsible was a twelve-year old Rottweiler-retriever cross that, due to a skin condition, had a shaven body but mane of hair around its head. According to the reputable (!) *Daily Star* of 17th July 1999, the sightings began when Rocky's owner, Joanne Story from Birdwell, had taken him for a walk at a nearby pond. An eye-witness named John McLoughlin reported,

> "I'd never seen anything like it before. It looked like some kind of lion or leopard. It had a furry face and bushy tail but no body fur."

Despite the case of mistaken identity, was Rocky also to blame for the sightings of a creature that was, according to witness Ray Cibor, "eight-foot long, yellow, black and orange with black stripes and a yellow patch at the front" which terrorised him at his farm in Armthorpe, near Doncaster in June of 1999? "I know a tiger when I see one," Ray told the *Daily Star*, but clearly he didn't!

In 1976 there was a lion scare in Nottingham after two witnesses, milkmen David Crowther and David Bentley claimed to have seen a cat from a distance of fifty yards that resembled a lion. The witnesses contacted the police who searched the area in a car and through loud-hailers instructed locals to stay indoors. According to author Graham J. McKewan 'Over the next eight days the police received some sixty-five reports of a lion in the area,' but most of these were explained as misidentifications of domestic cats and large dogs but this still left the authorities with a small percentage of eye witness reports that seemed to speak of a large cat-like animal. With the police out in full force it was no surprise that young children took to the woods in search of the beast but on August 6th of 1976 the hunt was called off despite only a few days previous Chief Inspector John Smith had commented that, "We are almost totally convinced it is there..."

Hairy hoaxes...
Judging by the numerous reports made over the years of so-called 'big cats' in the wilds of Britain, an alarming number of witnesses do not know what they are seeing. It's also fair to say that a number of police, welfare officers and even 'big cat' researchers don't know what animals do and do not roam the UK woods. In September 2012, I received a report from a man who claimed that whilst driving through a wooded area of Kent he'd seen a lion run across in front of his vehicle. Then I heard on the radio another 'big cat' researcher popping up and saying that "there's been reports across the UK of lions, tigers, jaguars and cheetahs" as if such cats are surviving without detection in our forests. It's no wonder 'big cat' sightings are merely headline fodder for newspapers during a quiet week. When police were called out to investigate a lion in 2001 in deepest, darkest Blagdon, in Somerset, you would've thought that they'd learned their lesson, and yet over a decade later the leafy suburbs of Essex are plagued once again by authorities wasting time and money looking for an impossible cat.

I have numerous cases on record of large cats escaping from captivity and police scouring fields and woods in search of, but I thought I'd leave you with my two favourite lion-related events from British history, both emerging from the county of Sussex. The first was recorded by several newspapers, including the *Waganui Chronicle* of May 8th 1905 which records 'Escaped Red Lion' after villagers throughout the south of England had barricaded themselves in their homes after reports of an escaped lion. The animal, allegedly sighted in Sussex as well as Hampshire, was thought to have belonged to a menagerie. A postman was said to have seen the cat chase and devour three sheep in the neighbourhood of Petersfield but the story reached hysterical proportions when it was claimed that the lion had 'eaten three school children at Harting,' even though no-one in the region could prove the rumour true. Others claimed that the beast had been shot dead at Didling but locals there stated categorically that the animal had been taken down in another village. After several weeks of panic the truth finally came out, and a truth far weirder than the recent Essex lion scare, with the Chronicle commenting that a Hampshire pub called the *Red Lion* had in fact lost its sign, and so locals began saying that, "The lion has escaped..." and the "lion had gone missing" and those not in the know of the

30 police officers, crack marksmen and two helicopters join search

HUNT FOR A 'LION ON LOOSE IN ESSEX'

Fearsome: A lion is believed to be running free in fields

ARMED police were last night hunting for a lion believed to be on the loose in Essex.

Residents were told to stay indoors after the beast was spotted in fields near Clacton. Two police helicopters and

By John Stevens and Hannah Roberts

around 30 police officers were scrambled after locals reported the sighting.

One witness described onlookers screaming as they realised they had just seen the predator.

Ruth Bailey 19, was walking with

his two boys, aged nine and 11. He said: 'A man started running towards us yelling "It's a lion". He looked so panicked you knew it was not a joke.

'The lion was in the middle of the field with its back towards me. When someone else screamed it turned around and

Turn to Page 5

pub locals jumped to conclusions and believed a real wild animal was on the rampage.

The last story, which pretty much sums up the Essex lion scare, took place, or so they say, in 1933 in Sussex and was mentioned in *The Times* of 19th October under the heading of 'Sussex lion hunt.' Four men were charged in a Sussex court with public mischief after they started a scare story that a lion, named Rex, had escaped into the wilds of Bognor Regis in the July of '33. William Edmund Butlin, Alan Leslie Proctor, Clifford Stanley Joste and John Waller, were said to have unlawfully performed an elaborate hoax in which they claimed that whilst a male lion was being transported in the vehicle of Butlin, it had escaped at Clymping and killed a sheep. Officers of the West Sussex Constabulary were called out to investigate the incident only to find that no lion had been sent via vehicle to or from Sussex and the scare was deemed a hoax. All four men involved in the prank were accused of wasting the time of the police, who, in being called out to investigate the false escapee, had according to the newspaper, deprived the public of their usual high standard of service.

And so, as you can see, in most cases of lions escaping into the British wilderness, they are quickly recaptured, shot dead or as in the case of the Essex beast, a non-event. This doesn't of course mean that exotic cats in general do not roam Britain, but the media are far hungrier for a dramatic headline, in the same way a lion would be for large prey and all the while there are people gullible enough to fall for such drama, these types of panics will always create a headline. Or, as one conspiracy theorist recently told me, "I think the Essex 'lion' hunt was created by the authorities to cover up the fact there had been a terrorist threat in rural Essex" and that my friends, just about sums it up!

Neil Arnold is a full-time researcher regarding 'big cat' sightings and evidence. He is the author of *Mystery Animals of the British Isles: London* and *Mystery Animals of the British Isles: Kent*. His websites are: www.kentbigcats.blogspot.com and www.beastsoflondon.blogspot.com

Le Gevaudan:
The Man behind the Monster

Paul Williams

Betwen 1764 and 1767 the harsh mountainous region of the Gevaudan in Northern France was terrorised by a beast that killed over a hundred people and injured many more, before being shot by a peasant named Jean Chastel. Its ability to evade the nation's finest hunters created a legend, immortalised in dozens of books and a successful movie *Brotherhood of the Wolf*. This enduring popularity is not just due to the unprecedented ferocity of the attacks and their exceptionally long duration but to the perceived failure of the authorities and later researchers to successfully identify the beast. Werewolves, hyenas, wolfdogs, prehistoric survivors, Tasmanian tigers, monkeys, tigers, and wolves were all proposed as candidates in perhaps the greatest zoological conundrum of all time.

The mystery was apparently solved in 1997 when a taxidermist named Franz Jullien uncovered evidence that a striped hyena had been on display in Paris, described as the beast. This was recorded in a booklet published in 1819, which listed some of the animals in the cabinet, an early museum, of natural history. Pages 5 and 6 of the booklet described the hyena as being a fierce animal in the same class as the wolf and lynx. As well as devouring corpses it attacked men, women, and children. It wore a mane like a tiger's on its back and was described as "the same species seen in the cabinet of natural history which ate in the Gevaudan a large number of people." The generic description originates in an encyclopaedia that predates the beast and appears in early and later copies of the booklet. The crucial last line indicates not only that a hyena was the beast but also that the body was preserved.

The theory has not been accepted by everyone but, by accounting for the body of the beast it raises questions about the man who killed it, Jean Chastel. It is over fifty years since an article in *Fate* magazine suggested that Chastel might have been a sorcerer. More recent research has increased speculation that events following the death of the beast were not as described.

On 19 June 1767 Chastel, a sixty year old peasant succeeded where three of the King's finest hunters, and their troops, had failed. He joined a hunt in the woods near Tenaziere and brought the beast down with a single shot. Triumphantly he displayed the body to locals for ten days then took it to Versailles where the King proclaimed his disgust at the stench of the decomposing corpse and refused to pay any reward.

Chastel did receive 72 livres from local authorities for his troubles, a not insignificant sum in those days except when compared with total rewards of over 6000 livres that had been offered for the beast's destruction. The King himself had authorised a payment of 300 livres to a boy who fought against the beast on 12 January 1765 and the same sum to the children who were present at the encounter.

Jullien's research indicates that the body was preserved not burnt or buried. This tallies with instructions from the court, as late as January 1767. Many natural history experts were members of the court and expressed an interest in identifying the famous beast. It is unlikely that they would have allowed its destruction. This preservation must have been before the body reached Versailles as the length of time taken by the journey would make it impossible to save. Therefore it would not have smelt and earned the King's displeasure. Why, then, was Chastel turned away and not rewarded by the court? There are several possibilities. The King had already rewarded the royal hunter, Antoine de Beauterne, who killed a large wolf on 21 September 1765. This animal was preserved and paraded as the beast, in ignorance of the continuing ravages in the Gevaudan. Antoine and, by default, the monarchy could not afford to be upstaged by a commoner.

During his hunts Antoine angered many of the locals with his tactics, such as leaving bodies of the dead overnight in case the beast returned and imprisoning those who failed to report attacks. In a society fast approaching revolution, he was resented by many of those he was trying to save. On a hot August day in 1766 two of his men, Louis Pelissier and Francois La Chesnaye, were riding through the forest at Montchauvert when they met Jean Chastel and two of his sons. The hunters asked if the land ahead was marshy and were told that it was not. Pelissier proceeded then sank into a deep bog and was only just able to save his horse. The Chastels laughed at his misfortune. Enraged Pelissier tried to assault the youngest Chastel only to find the others pointing their guns at him. When Antoine heard of this incident he arranged for all three Chastels to spend several days in prison. This injustice, it was said by some, inspired Chastel to seek revenge with his magical powers.

It was believed by many at the time that the beast had a supernatural dimension, able to inflict damages far greater than anything ever done by wolves and to escape without leaving a trace. Stories told of it jumping over high walls and, on one occasion, even talking. The isolated Gevaudan area was rich in werewolf folklore and just a few centuries earlier several Frenchmen and women had been executed for the crime of lycanthropy or, to be precise, for eating children whilst in the guise of a wolf.

Eating was used as a metaphor for sexual crimes. This is seen in the *Red Riding Hood* parable of seduction or rape, a French tale first written down in 1697 from earlier oral versions. There was also a curious belief that werewolves undressed female victims. No adult males were killed by the beast. Modern analysis of some of the lycanthropy trials suggests that the horrific crimes described may actually have been sexual attacks on children. During the rage of the beast more incidents than those documented were alleged. One of Chastel's sons, Antoine, disappeared after the death of the beast, leading to suggestions that some of the attacks were faked to disguise his sexual offences against children. It is interesting that Jean and Antoine Chastel appear on official records as the witness to some of the deaths.

In 1936 the French novelist Abel Chevalley alleged that Antoine Chastel kept a pet hyena in his menagerie. Forty years later the naturalist, Gerald Menatory, suggested that Antoine had obtained the hyena when in Africa as a boy and subsequently trained it. Menatory went on to found a wolf park in the Gevaudan and wrote partly to exonerate the wolf. His book led to rumours that the death of the beast was staged with the hyena following orders from its master.

As both Chevalley and Menatory wrote before Jullien identified the beast the theory should be looked at again. Hyenas, like wolves, do not habitually attack people and the ravages of the beast have no parallel in recorded history. Moreover the stripped hyena is considered less dangerous than its spotted cousin. Man-eating hyenas have been known in India, Malawi, Mozambique, and the countries now known as Azerbaijan and Turkmenistan. One of the Indian incidents extended over a three year period but with fewer attacks than in the Gevaudan. The anomaly could be explained if the beast was trained to be more ferocious than normal. The beast's ability to avoid hunters would seem less extraordinary if its lair was not in the open but, safe from the hounds, in captivity.

The situation is further complicated by the existence of other man-eating beasts in surrounding regions of France before and during this period. These appear to have been hybrids of wolves and dogs. The beast of Sarlat, killed in July 1766, had some physical characteristics of foxes and greyhounds. There were also signs of hybridization in the wolves of Perigord, killed in February 1766 after killing eighteen people. King Louis XIV also took a personal interest in this case and, like Sarlat, it is within range of the Gevaudan for a wolf's territory.

The climate was right then, for a manufactured beast to emerge and take the blame for all manner of crimes. It is also possible that Antoine's claims to have killed the real beast were justified and that Chastel used the trained hyena to discredit this claim. A somewhat extreme method of extracting revenge on the man who jailed him but life was cheap in those times and the promise of 6000 livres would tempt more audacious crimes. When Antoine's reputation remained undiminished the only option left to the Chastels was to publicly eliminate the wolf and claim the reward money. That this too met in failure reflects more the soon to be abolished hierarchical society than any royal knowledge of deception.

If Jean Chastel knew the secret of the beast he took it to his grave. The gun used was later sold to Abbe Pierre Pourcher, the man who first documented the case at the end of the nineteenth century. Pourcher, like many of those at the time, believed that the beast was sent by divine intervention. Human intervention seems more likely and perhaps, when Jean Chastel fired the bullet, he was atoning for the sins of his son.

2012: A Year in the Life of the Centre for Fortean Zoology

I have always wanted the CFZ to be a proper family, and in many ways I have achieved this aim. However, the thing about families is that bad things happen to them as well as good ones. Whereas I would like the CFZ to be a storybook family as described by Enid Blyton in which we all sit about drinking ginger beer and having a jolly good time, sadly things don't always work out like that.

In recent months both Nick Redfern who runs the CFZ United States office, and Tony Lucas, who runs the New Zealand office have had serious personal issues with which to deal. They both assure me that they will be back in the saddle, but for 2012 there are no annual reports from them.

CFZ 2012 Report
by Rebecca Lang and Mike Williams

H ere we are again, reminiscing about another year of cryptozoology and general weirdness Down Under. It's only when you look back on the year that you realise a hell of a lot actually happened. And so here are the highlights, and some lowlights, as we saw them:

Thylacine skulls, searches and mainland sightings

January 2012 kicked off with an exciting development – the discovery of an alleged Thylacine skull by some trail bike riders. Sadly the skull's discoverers - brothers Levi and Jarom Triffitt – were quickly disabused of their original verdict.

> "The museum have told us its a dog skull but still can't identify what sort, our story is as we have filmed it bar the lobster which was filmed 2 months ago, however we were looking for lobsters and to find this skull was an amazing experience we'll never forget, we will still live in hope they are out there somewhere..."

That same month a couple of Australian academics said people should stop wasting time and money looking for the Tasmanian tiger.

Dr Diana Fisher and Dr Simon Blomberg from the University of Queensland's school of biological sciences said since the last wild thylacine was captured in 1933, there had been ongoing searches and numerous unconfirmed sightings of the carnivorous marsupial. But, says Fisher, such efforts were misguided.

> "There's been more search efforts for the thylacine than any other mammal globally," she says. "I think that's just a waste of money."

And in February, sightings of an unusual 'dog-like' animal, seen in and around the region of northern NSW, piqued the interest of north coast wildlife expert and cryptozoologist Gary Opit.

For the past 15 years, Mr Opit has been a regular guest speaker on ABC North Coast (formerly 2NR) and in that time he has received around 50 reports from callers describing an unfamiliar animal that he said resembles one of two species, possibly the marsupial lion (*Thylacoleo carnifex*), or the Tasmanian tiger (*Thylacinus cynocephalus*).

Many of the recent sightings have occurred around Mullumbimby, Nimbin and Byron Bay. The idea of mainland thylacines isn't so far-fetched – they once roamed all over the Australian continent.

Vale Chief – White Wolf of the West
Our beautiful crypto dog, Chief, a white german shepherd, shuffled off the mortal coil in February, a definite lowlight - not just for us but for all the people who have crossed our paths over the years and stopped to ruffle our 'little' man's fur. Loved by all, Chief was our constant companion and joined us on countless adventures chasing reports of big cats, yowies and all manner of crypto beasts.

He was a well-known face in our town, and even earned the nickname The White Wolf from some neighbours and council workers who found him every bit as terrifying as White Fang! He will be sadly missed.

Black fox sighting
In March a Victorian woman, Diane, living near Ballarat reported having seen a black fox on her way to Beaufort from Ballarat.

She commented:

> "I saw the fox when I was driving out to my sisters out at beaufort, it ran out right in front of me, it didn't have the white tip on the tail though. It was last Tuesday 6th March 2012 at approximately 6pm. It was just after Trawalla."

While a very rare animal in Australia, black foxes are nonetheless a known part of Australia's introduced fauna; being a melanistic version of the red fox, known colloquially as a darker variant of the silver fox. These animals have been reported in Australia since the 1920s and on some occasions have been mistaken for melanistic leopards (panthers) and occasionally described as a melanistic puma. One particularly pertinent example was a "Black Panther" reported shot by two youths near Frankston that was later described as a large Black Fox upon investigation by police.

In our *book Australian Big Cats: An Unnatural History of Panthers* we raised the issue of rare Black Foxes being mistaken for dark coloured big cats. Rebecca described, in particular, an incident where she saw a quadruped which, while moving with a somewhat cat like grace, was clearly neither a cat nor a dog. The book also refers to a University of Western Australia Fox

DNA project that received pictures of a black fox shot and brought in for investigation on a property in South Australia. Our thanks to David Waldron for this report.

Victorian big cat search called off

An official search for proof of the Victorian big cat was called off in May 2012. Department of Primary Industries staff assigned to a state-wide hunt were redirected and Agriculture Minister Peter Walsh told the media the program had been shelved. Mr Walsh was asked about the search for the big black cat during a budget estimates hearing at Parliament yesterday.

> "It's something that has been a part of Victorian folklore for quite a few decades," he said. "There are those who firmly believe there are some big cats out there. There are those who are equally as certain there are no big cats. When the resources are available, we will have a search of all the information about the big black cat issue."

Yowie Tales – photographs of yowies?

In May 2012 a book, the first in a projected series by Queensland researcher Brett Green, was released purporting to show never-before-published photographs of the Yowie, Australia's very own Bigfoot.

Brett Green declined to answer any questions put by the CFZ about the photographs or his book. The two images of 'yowies' in the book, if ever proven to be genuine, could be some of the most important secondary evidence of unusual creatures ever recorded. Unfortunately his reticence to answer any questions about the matter sheds considerable doubt on their veracity.

Magnetic Island panther returns

Back in September 2011, CFZ Australia was contacted by Sabrina Tessari, an Italian tourist who had visited the tourist mecca Magnetic Island. She emailed us with some very interesting news.

On her visit Sabrina was keen to see the local wildlife, but she was shocked and surprised to see a big black cat on the island - an animal "equal to a black panther". At the time even we were puzzled but open-minded. And now subsequent reports have everyone scratching their heads.

Magnetic Island sits 8km off the Australian coast and smack bang in the middle of the islands of the Great Barrier Reef off northern Queensland. For animals that characteristically don't like water, one has to wonder how this particular panther arrived at its destination!

The panther returned in June 2012, much to the surprise of residents – many of whom were initially skeptical about Ms Tessari's claims.

On June 4, around 8.30pm, under the light of a full moon on Monday night, Magnetic Island resident since 1973, Patti Winn was walking home when large branches in a nearby tree began shaking and swaying wildly.

"There was something hefty moving in the branches and as I moved away across the road it moved into another tree (behind the first). I looked back and saw a body form coming down the trunk. I didn't see the head or a tail but it was as big as a medium sized dog. I panicked a bit. I knew it wasn't a possum. I just ran home. I just wanted to get inside! I don't know where it went."

QPWS Ranger Partick Centurino said another sighting was made at the end of Mandalay Avenue: "A woman who was cycling saw it. She was stopping at the entrance to the Hideaway Estate and could see it next to the entrance to the walking track. It was all black and as big as a small Labrador with a cylindrical round tail." The plot thickens!

Yowie man footage knocked on the head
We were pleased to clear up another supposed yowie mystery that cropped up in the tabloid press in June – that of a thermal camera image purporting to show a yowie in the Blue Mountains of NSW, not too far from our own digs.

Previous online analysis by the CFZ's Mike Williams had already shown that the image was, in fact, of a possum on top of a rock! The thermal signature of the 'yowie' was the same as the surrounding trees and the rock - mystery solved! The story aired in the wake of a visiting US television show, which claimed to have recorded evidence of a yowie bellowing in remote bushland, and generated a small amount of interest in the tabloid press.

Victorian big cat search backflip and verdict
One minute it was off, then it was all systems go again! Just a couple of months after it shelved plans to solve Victoria's 150-year-old big cat mystery, the Victorian Government decided in August to put the big cat issue to rest once and for all.

Witness sightings have been logged for the past 60 years by government agencies and private researchers of cougars, panthers or pumas from Gippsland to the Otways, the Grampians, central Victoria and at Beechworth in the northeast.

The Victorian Government ended its 'desktop study' into the existence of big cats in Victoria just a few weeks later in September, declaring it was "highly unlikely" the animals existed, and that instead, they were probably large feral domestic cats.

"The big cat study was done with existing resources of the DPI and the DSE, the majority of work was done by researchers at the Arthur Rylah Institute. There is not an exact figure," Victorian Agriculture Minister Peter Walsh said.

Of course we thought it was a bit rich – a desktop study that ignored much of the best evidence collated in the past 10 years!

So of course we addressed all of the points in the report on the CFZ Australia blog. Rather than bore the pants off everyone here, the full analysis can be read at: http://www.cfzaustralia.com/2012/09/the-big-cat-report-that-wasnt.html

It was with a touch of irony that earlier that same month a Victorian zookeeper had her own big cat sighting. Previously a skeptic, the shaken woman said "It was the size our dingoes, it was on all four legs and it moved like a cat and it had a long tail coming after it. I know it sounds crazy and I wouldn't have believed myself." The animal bolted past her through an open gate, before she ran to hide in a nearby reptile shed. "I was so scared, thinking I was going to die."

New Victorian big cats book

Following hot on the heels of all these big cat shenanigans was the November launch of David Waldron and Simon Townsend's book *Snarls From The Tea-tree: Victoria's Big Cat Folklore*. The academic book examines how the 'myth' of the big cat has evolved over time, but relates closely to Australia's engagement with its environment. The book is available from Australian Scholarly Publishing.

New thylacine book on the horizon

Veteran thylacine researcher Col Bailey will release his long-awaited autobiographical tome on his search for the extinct marsupial wolf in May 2013 through Five Mile Press. Col's book details, among other things, his own sighting in southwest Tasmania's rugged bushland on an expedition in 1995.

He said the officially-extinct carnivore appeared while he was camping alone in the Weld Valley and he followed it beyond a cluster of ferns. "It was about 15 feet away and it turned and looked at me for several seconds, then backed away," he said. Mr Bailey said it stopped and looked at him a second time before it disappeared into the scrub.

Looking ahead

What cryptozoological adventures lie ahead?

You can keep abreast of what we're up to at www.cfzaustralia.com

Until next year!

CFZ Canada 2012

Robin Bellamy

Although reports of unusual animals have been rare in Canada in 2012, it has still been a year of discovery. Recognizing the who, what, where, when, and how of research in general and the unknown animals specifically has been a journey this year. No good researcher ever rests. When there is little field work it is an opportunity to learn more about not only the study in general, but also ourselves as researchers.

This "Year of the Dragon" kicked off with a bit of a partnership with the Royal Canadian Mounted Police. Western detachments were very receptive and helpful when approached for information and opinion on their local Sasquatch reports. We learned, among other things, that each report is investigated. It is never assumed that the witness is lying or impaired in some way. This is comforting, as BC is a hotbed of Sasquatch sightings every year. It is good to know that law enforcement is open to possibilities. Any reports that seem to have merit enough for follow up is generally passed to the local police or conservation authorities. After well over a century of witnesses being admonished, it is most pleasing to know that if you experience a Sasquatch you will be treated with some respect.

This year has also been sobering for those in this field. The tragic death of Jeff Rice, one of the people associated with the production of SyFy's Destination Truth series. Although not onsite with an investigative team and subject to the dangers of wild animals (known or unknown), this supportive role in research found him in Uganda and vulnerable to some local savagery. CFZ Canada made a concerted effort to warn investigators of the dangers of investigation and suggested items to bring along. We also lost Jeanett Thomas, the wife of Lars Thomas who is a significant contributor to the study in Denmark. Jeanett had been missing for about six months before they found her, a victim of what is probably a tragic accidental drowning. Then in November a young ghost researcher died of a virus in North Carolina. Sara Harris contracted the virus during an investigation in an abandoned home in which there were copious amounts of bat and rat droppings. It is evident that we as researchers need to take safety more seriously, and CFZ Canada has made a commitment to speak to this issue more regularly.

While researching in libraries and online is considerably safer, it is no less exciting. This year we looked at some of Canada's water creatures. From an underwater feline in the Great Lakes region to a strange "Bat Signal" shaped living creature off the coast of British Columbia, we were swimming with reports and study on sea monsters and weird wet animals. There were the typical bloated carcasses of course, but more intriguing were a "metrosaurus" and mermaids. Some lesser known cryptids that we learned about were werewolves (loup garou) and Mothman. While not common in Canada, these types of creatures do come to visit from time to time and make it into local lore or newspapers. Currently, what is known as the "Michigan DogMan" has seen resurgence in southwestern Ontario and may be a similar creature to Quebec's legendary wolf man.

The big research project for 2012 was an analysis of Sasquatch. As he is by far the most common Fortean zoological being seen in Canada, we took several weeks to deeply research the similarities of these creatures to both humans and other primates. "Sasquatch from the Bottom Up" is a seven article series which includes information and comparison on everything from brain size to the series of arches present in the print castings. Late in the year, the world was abuzz with rumor that a DNA genome had been discovered at a lab in the US and that "Bigfoot" would finally be defined and given biological status. This study is awaiting publication and peer review, however, and until the claims are substantiated we will continue to roam the woods in search of our own evidence.

Finally, we took a more whimsical look at what is generally a very serious scientific study. With an article on faeries and another on an smart phone application designed as sort of a cryptozoological scavenger hunt, we sometimes weren't quite so "academic". Moving forward we hope to continue to add the occasional smile, but remain dedicated to a better, safer, and more thorough study of creatures unknown.

BFF Annual Report 2012

1st Quarter
Jan-Mar 2012:

- The BFF Forum Management Team secured independent funding and transferred operations to a new server with considerably more space for expansion. Thanks to Jon Downes and Graham Inglis for their support and assistance during this time.

- The BFF Forum Management Team welcomed (Author) Kathy Strain as the newly appointed Steering Committee Chair person.

- The FMT/Steering Committee completed deliberations on the new BFF Premium Access Forum area; the area was completed and included the addition of the BFF 1.0 archives.

2nd Quarter
Apr-Jun 2012:

- Several glitches with the new server were identified and resolved by the BFF tech "gigantor".

- Premium memberships were instituted and the BFF signed 48 subscriptions this quarter.

3rd Quarter
Jul-Sep 2012:

- After several back and forth correspondence with the former domain holder, The BFF procured the "Bigfootforums.com" domain name and licensed it to the forums proper.

- After months of unsuccessfully trying to obtain forum software updates through the previous license holder, The BFF FMT obtained an independent license through the same company so we could continue to operate safely and effectively.

- The BFF signed <u>31</u> Premium Membership subscriptions this quarter

4th Quarter
<u>Oct-Dec 2012:</u>

- The BFF FMT welcomed a new Chief Administrator "See-Teh-Cah NC" who will oversee staff operations.
- Premium Membership continues to increase. The current total for 2012 is <u>90.</u>

This report was compiled by:
masterbarber
Director, Bigfoot Forums

Dear Friends,

This is the 20th time I have sat down to write a CFZ Annual Report. It makes me feel old.

The CFZ is currently in the state of flux, but looking back over the past two decades it seems that we always were. Ten years ago George Harrison, my favourite Beatle died of cancer.

His posthumous album contains a song which includes the words: "if you don't know where you're going, any road will take you there", and if I hadn't already chosen a Latin tag *Pro bona causa facimus* (we do it for a good reason), which I pinched from a children's book called *The Case of the Silver Egg* by the late Desmond Skirrow, then I would probably have adopted George Harrison's words. It is interesting, by the way, that I have never managed to find any other reference to this Latin motto. I had assumed that it was Scipio, Cicero or one of those dudes from ancient Rome that I learned about during my Latin classes in Bideford Grammar School all those years ago, but on the Internet the only reference I can find at all are in things written by me.

I only just realised, literally whilst typing these words that my choice of motto could well be seen as quite significant in that it tells the story of a group of children living in 1960s London who get involved in a major international espionage mystery, and come out on top. The important thing about the story is that the Queen Street Gang do things in their own way, and usually without adult interference or supervision. Substitute the Queen Street Gang for the CFZ, and substitute 'adults' for erm... the scientific establishment, the established media, and pretty well anybody else you can think of with whom we have come in contact over the past two decades, and you have fairly good encapsulation of the ethos of the CFZ.

When I look back over the last few years I see a CFZ that has changed rapidly in a very short length of time. And it is changing now, very much so. For example, when we first moved to North Devon, we expanded the CFZ animal collection greatly. In fact, with hindsight, we expanded it too much. Now we are downsizing, and we're doing so for a very important reason. When we first came here we intended to build a museum at the top of the grounds. We still have a small but interesting collection of objects from our various investigations, but now we have decided that it is far more important to have input into far more extensive exhibitions elsewhere. For example, in October/November 2012 we had a very successful exhibition at Barnstaple museum, due to the kind offices of Julian Vayne.

As Corinna and I get older I realise that the way that we lived only a few years ago, sharing our living space with an ever-changing ménage of people is something which a couple well into their fifties really cannot do any more. We still have visitors, and my old friend Richard Freeman, who like Graham Inglis - who still lives with us - is more like a brother than a friend, is still a frequent visitor. However, rather than looking for people to come and live with us, we now look for volunteers who are happy to give up some of their time and expertise on a regular basis to help further the CFZ ideal.

I am particularly proud of our achievements in the world of publishing. During 2012 we published our first full-colour book, and also sponsored, financed and published *The Journal of Cryptozoology*, the first English language peer-reviewed cryptozoological journal since the glory days of the ISC several decades ago. The editor, Dr Karl Shuker wrote:

> Welcome to the *Journal of Cryptozoology*. Following the demise of *Cryptozoology* (published by the now-defunct International Society of Cryptozoology), there has been no peer-reviewed scientific journal devoted to cryptozoology for quite some time. Consequently, the Journal of Cryptozoology has been launched to remedy this situation and fill a notable gap in the literature of cryptids and their investigation. For although some mainstream zoological journals are beginning to show slightly less reluctance than before to publish papers with a cryptozoological theme, it is still by no means an easy task for such papers to gain acceptance, and, as a result, potentially significant, serious contributions to the subject are not receiving the scientific attention that they deserve. Now, however, they have a journal of their own once again, and one that adheres to the same high standards for publication as mainstream zoological periodicals.

I think that when someone takes a look at our achievements, maybe in 20 years time when we will have been in operation for 40 years, and I - if I'm still alive - will be well into my dotage, they will find that our greatest achievement will have been in publishing a long list of books which needed to be published, but which nobody else would have touched with the proverbial bargepole.

PUBLISHING
We published 24 titles this year in six different imprints. They were:

CFZ PRESS
Centre for Fortean Zoology Yearbook 2012 edited by Downes, Jonathan
The Mystery Animals of Pennsylvania by Gable, Andrew
SEA SERPENT CARCASSES: Scotland - from The Stronsa Monster to Loch Ness by Vaudrey, Glen

Wildman! by Redfern, Nick
Globsters by Newton, Michael
Cats of Magic, Mythology and Mystery by Shuker, Karl P. N
Those Amazing Newfoundland Dogs by Bondeson, Jan
CFZ - 1992-2012: The Thoughts of Chairman Jon by Downes, Jonathan

FORTEAN WORDS
The Grail by Coghlan, Ronan
HAUNTED SKIES Volume Four by Hanson, John and Holloway, Dawn
Quest for the Hexham Heads by Screeton, Paul
UFO WARMINSTER: Cradle of Contact by Goodman, Kevin
HAUNTED SKIES Volume Five by Hanson, John and Holloway, Dawn
HAUNTED SKIES Volume Six by Hanson, John and Holloway, Dawn

FORTEAN FICTION
Left Behind by Wadham, Harriet
Snap by Bredice, Steven
Green Unpleasant Land by Freeman, Richard
Dark Ness by Cope, Tabitca
Death on Dartmoor by Francis, Di
Hyakumonogatari Book One by Freeman, Richard
Dark Wear by Cope, Tabitca

CFZ CLASSICS
Head Hunters of the Amazon (Annotated edition) by Up De Graff, Fritz W

JOURNAL OF CRYPTOZOOLOGY
THE JOURNAL OF CRYPTOZOOLOGY: Volume One edited by Shuker, Karl P.N

CFZ COMMUNICATIONS
LANDMARK NORTHAM - Bone Hill: Northam's Best Kept Secret by Jackson, Jim

We have had a few problems this year with ex-clients. From the beginning I have always tried to run the publishing much after the manner of the late Tony Wilson and Factory Records. It has worked fine as long as everyone involved behaves like a gentleman. Unfortunately, this year, not everyone has. We will be re-jigging the contracts to cover us from these unfortunate occurrences ever happening again. We also now have the facility to print books in colour and to produce ebooks, so the new contracts will also reflect these new opportunities for us all.

WEIRD WEEKEND 2012
 This year's event was once again held on the third weekend of August. The Programme was as follows:

FRIDAY
Richard Freeman: 20 Cryptids you have never heard of

Paul Screeton: The Hexham Heads
BOOK LAUNCH: Quest for the Hexham Heads
Richard Thorns: The search for the Pink Headed Duck
Bedtime Story with Silas Hawkins
SATURDAY
Intro to Cryptozoology
Nick Wadham: The Bugfest Deadly Animal Show
Max Blake: An analysis of the Borley Bug
Harriet Wadham: Book Signing
Jonathan McGowan: Large cats in Britain - The Dorset enigma
Glen Vaudrey: Scottish sea monster carcasses
BOOK LAUNCH: Scottish sea monster carcasses
Jan Bondeson: Greyfriars Bobby
CFZ Awards
Film Premiere: Heads (Dir. Graham Williamson)
The Hexham Heads - Q&A with Graham Williamson and Paul Screeton
Bedtime Story with Silas Hawkins
SUNDAY
Richard Muirhead: The Flying Snake of Namibia
Lars Thomas: Danish Cryptozoology
Richard Freeman et al: Sumatra 2011
Lars and Jon: Wild Woolsery - the results of the Nature Walk
Ronan Coghlan: Sinbad the Sailor
Jon Downes: Keynote Speech

Speaker's Dinner at the Community Centre
Film: Occasional Monsters
Silas Hawkins: Final Bedtime Story

I would like to say a big thank you to Matthew and Emma Osborne for their hard work both for the Weird Weekend and for other projects during the year. We really couldn't manage without you. I would like to thank everyone at the Community Centre especially Simon and Sharon Bennett for everything they did for us. Your kindness humbles me.

ANIMALS
The most important news this year concerns our Rio Cauca caecilians *(Typlonectes natans)*. We were proud to breed them in 2011, and devastated when all four babies died in the winter. However, we bred them for the second year running this year. They are in a new tank at a slightly higher temperature and with a wide range of diet. If they survive the winter, two of the babies will be going to London Zoo in the spring.

We have also acquired a single chubby frog, and hope to be building up colonies of this species. The village pub has changed hands and no longer wants to be involved with our outreach programme so we have removed our fish from there.

FUTURE

Now, what does the future hold for us? Of course, I cannot answer that question with any degree of certainty, and bearing in mind quite how many of what Graham calls curveballs have come our way over the past two decades, I would be even more foolish than normal if I tried to foresee the future. However, what I *can* do is to tell you what I would like to see happen.

I created the CFZ in my own image (and by this, I mean that I created it according to an image in my mind rather than having created it to look like an ageing fat hippy) and to a greater or less extent have been steering it in my desired direction ever since. It is interesting that when, like I have over the last few days, you look back of the history of the organisation you can see that it has had several distinct phases.

PHASE ONE: the infancy, during which it was run purely by me and my first wife Alison. (1992 to 1993)
PHASE TWO: our first phase of expansion during which Alison and I were joined by the late Jan Williams, and we started publishing. (1994 to 1996)
PHASE THREE: in the immediate aftermath of Alison's and my divorce, I was joined first by Graham, and then by Richard, and the three of us managed the running of the organisation quite happily, although at this stage we were still basically a theoretical and publishing organisation. (1996 to 2002)
PHASE FOUR: as my mental and physical health improved, and - in the wake of my mother's death - my income also improved, we began to do more and more fieldwork, and to publish more books (2002 to 2005).
PHASE FIVE: after the move to North Devon and then in the wake of my father's death when I was able to divert considerably more funds into an ambitious campaign of publications. Things became even more coalesced in 2007 when I married someone who took it upon herself to make the administration of the organisation work properly for the first-time.

Now we are approaching phase six.

My family arrived in the little village of Woolsery in July 1971 and I soon became friends with the three children who lived next door. David was about the same age as me, and to my great joy I found out that we were soul-mates. Like me he had a voracious appetite for knowledge about the natural world and its denizens, like me he read everything that he could get hold of, and like me he had a surreal and slightly peculiar sense of humour. We soon became very close friends indeed, and were inseparable during our schooldays. He was sceptical - though interested - in my newfound passion for cryptozoology, and together we roamed the fields and woods, and investigated the local streams and ponds in search of the wildlife to be found therein.

His two sisters were younger than me: Lorraine by about two years, and Kaye by about four. David and I used to tease them unmercifully. Whilst they, too, were fond of animals, they found the wriggly things that their brother and I used to keep in jam jars in the garden shed to be uniformly icky. David and I realised this and used to torment the poor girls with biscuit tins full of spiders, and jam jars containing large and ugly horseleeches. It is - I believe - a testament to their good nature that we are still friends more than 30 years later. But in a situation analogous to that I described earlier regarding Richard Freeman and Graham Inglis, Lorraine and Kaye have always been more like sisters to me than friends, and especially since David died tragically young in 1987. Kaye has always referred me as her adopted brother and I have been Uncle Jon to her three children from the age when they first were self-aware enough to realise that the fat hippy in the corner was really quite nice.

A few months after I brought the CFZ to North Devon to help me look after my dying father, I telephoned Kaye. I was just being a nice Uncle, but I wondered whether her eldest son David (then aged 13) would like to come up and work for us on Saturday afternoons for a few quid here and there. He soon became invaluable, and after the second Weird Weekend he did with us I realised that we really would not be able to function without him. I don't think I treated him like a child after the age of 14, and at the time of writing he is approaching his 21st birthday, and I have only just realised that he is the same age as the CFZ - he has shouldered the responsibility of being my chosen heir with equanimity.

In 2001 I codified the structure of the CFZ formerly for the first time. At the top is our titular Life President John Blashford-Snell who took over the position when the previous incumbent Professor Bernard Heuvelmans - usually known as the 'Father of Cryptozoology' - died.

Below him was a three-person committee consisting of myself, Graham, and Richard. Being three of us it means that we never have a hung vote. However, I would like to say that in the ten years that we have been operating in this manner we have never had a serious disagreement. Below us there is what I dubbed The Permanent Directorate, and another group called the Advisory Board. The Permanent Directorate included people from the various study groups, the various international offices and those who have particular skills to offer. The Advisory Board are exactly what they sound like - a group of people who have particular expertise for knowledge in one specific area.

Now, ten years after setting this hierarchical management structure into place I am making the first major change. The three-man management committee has been - as of the 2011/2 management meeting - replaced by a five-person committee. Until now the only people eligible to vote have been me, Graham and Richard. Now, because of their invaluable contributions, I have expanded the committee to include my wife Corinna, and my nephew David Braund-Phillips. The Permanent Directorate, and the Advisory Board will still be there to advise and assist, but will not be able to make decisions.

I intend to continue our programme of publications, and in 2012 I have instituted a new series imprint called *CFZ Classics*. These will not just bring books that have been long out of print, and which are really only available to the cognoscenti, into the wider public consciousness, but will serve another, and equally important, function. There are people within the cryptozoological research community who have little or no income beyond state benefits. The last few governments have progressively demonised benefit claimants until it looks quite possible that we will see the end of state benefits as we know them in the UK within the next few years.

We have the technology and infrastructure available to publish as many books as we want, and the Internet is an invaluable marketing tool. Each of the books in the series will have extra essays, footnotes and as much additional material as we can provide. As each of the books in question is well out of copyright, the author's royalties will be paid to the person who put the package together. We will provide an unprecedented level of help to put currently impecunious researchers into the position where they can earn themselves a monthly income through their own efforts, and thus be able to lift themselves out of the poverty trap.

As I have written elsewhere, I had an unhappy childhood, and though I was a mildly gifted child, my family and teachers did their best to stifle my creativity and aspirations. Despite them I achieved most of what I wanted to, and am now in the position to help another generation of writers, artists, and dreamers. I won't embarrass them by naming names, but there are various people now in the scientific and cryptozoological establishment who have become what they are today, at least in part because I and the CFZ encouraged them when it mattered most. Ever since we moved to North Devon we have had more and more children becoming involved in what we do, and I think that this is massively important. We intend to do all we can to encourage literacy, and a love of nature, as well as encouraging the innate curiosity of succeeding generations of young people for as long as we can.

I have always believed that we are a family, in a very real sense, and now we are rapidly becoming a truly global family. And like a real family in the last year we have had several deaths, at least one birth, and human tragedy and triumph of various scales. Last year we raised money for the family of one of our longest serving Sumatran guides, Sahar Dimus who died suddenly, and we have done similar things as well. I hope that this is only the beginning, and that we can eventually run programmes all over the world to help the members of the CFZ family who are less fortunate than ourselves.

This year we had two young ladies spend work placements with us; Saskia England (aged 14) in the spring and Sheri Myler (aged 20) in the autumn. I hope that both of them enjoyed their placements with us and took away something of value from their time with us. We certainly learned a lot from both of them and hope to have many more trainees in the future.

Over the years I have made some bad decisions, and I have made some wrong decisions. My decisions to run a little museum and zoo in my back garden ultimately proved to be unwise, for example. However, over the past two decades the CFZ has done pretty well under my stewardship, and I am proud of what we have achieved together. I hope that whatever happens, we continue to be essentially a caring organisation, one that puts people before money, and common sense before ideology. I hope that the CFZ never loses its sense of humour, its sense of idealism, and never loses touch with its core concept, that half a century or more after Bernard Heuvelmans first brokered the ideal: the great days of zoology are not done.

Jon Downes,   CFZ, North Devon, 23rd December 2012

STILL ON THE TRACK OF UNKNOWN ANIMALS

The Centre for Fortean Zoology, or CFZ, is a non profit-making organisation founded in 1992 with the aim of being a clearing house for information, and coordinating research into mystery animals around the world.

We also study out of place animals, rare and aberrant animal behaviour, and Zooform Phenomena; little-understood "things" that appear to be animals, but which are in fact nothing of the sort, and not even alive (at least in the way we understand the term).

Not only are we the biggest organisation of our type in the world, but - or so we like to think - we are the best. We are certainly the only truly global cryptozoological research organisation, and we carry out our investigations using a strictly scientific set of guidelines. We are expanding all the time and looking to recruit new members to help us in our research into mysterious animals and strange creatures across the globe.

Why should you join us? Because, if you are genuinely interested in trying to solve the last great mysteries of Mother Nature, there is nobody better than us with whom to do it.

Members get a four-issue subscription to our journal *Animals & Men*. Each issue contains nearly 100 pages packed with news, articles, letters, research papers, field reports, and even a gossip column! The magazine is Royal Octavo in format with a full colour cover. You also have access to one of the world's largest collections of resource material dealing with cryptozoology and allied disciplines, and people from the CFZ membership regularly take part in fieldwork and expeditions around the world.

The CFZ is managed by a three-man board of trustees, with a non-profit making trust registered with HM Government Stamp Office. The board of trustees is supported by a Permanent Directorate of full and part-time staff, and advised by a Consultancy Board of specialists - many of whom are world-renowned experts in their particular field. We have regional representatives across the UK, the USA, and many other parts of the world, and are affiliated with

You'll find that the people at the CFZ are friendly and approachable. We have a thriving forum on the website which is the hub of an ever-growing electronic community. You will soon find your feet. Many members of the CFZ Permanent Directorate started off as ordinary members, and now work full-time chasing monsters around the world.

Write to us, e-mail us, or telephone us. The list of future projects on the website is not exhaustive. If you have a good idea for an investigation, please tell us. We may well be able to help.

We are always looking for volunteers to join us. If you see a project that interests you, do not hesitate to get in touch with us. Under certain circumstances we can help provide funding for your trip. If you look on the future projects section of the website, you can see some of the projects that we have pencilled in for the next few years.

In 2003 and 2004 we sent three-man expeditions to Sumatra looking for Orang-Pendek - a semi-legendary bipedal ape. The same three went to Mongolia in 2005. All three members started off merely subscribers to the CFZ magazine. Next time it could be you!

We have no magic sources of income. All our funds come from donations, membership fees, and sales of our publications and merchandise. We are always looking for corporate sponsorship, and other sources of revenue. If you have any ideas for fund-raising please let us know. However, unlike other cryptozoological organisations in the past, we do not live in an intellectual ivory tower. We are not afraid to get our hands dirty, and furthermore we are not one of those organisations where the membership have to raise money so that a privileged few can go on expensive foreign trips. Our research teams, both in the UK and abroad, consist of a mixture of experienced and inexperienced personnel. We are truly a community, and work on the premise that the benefits of CFZ membership are open to all.

Reports of our investigations are published on our website as soon as they are available. Preliminary reports are posted within days of the project finishing.

Each year we publish a 200 page yearbook

We have a thriving YouTube channel, CFZtv, which has well over two hundred self-made documentaries, lecture appearances, and episodes of our monthly webTV show. We have a daily online magazine, which has over a million hits each year.

Each year since 2000 we have held our annual convention - the Weird Weekend. It is three days of lectures, workshops, and excursions. But most importantly it is a chance for members of the CFZ to meet each other, and to talk with the members of the permanent directorate in a relaxed and informal setting and preferably with a pint of beer in one hand. Since 2006 - the Weird Weekend has been bigger and better and held on the third weekend in August in the idyllic rural location of Woolsery in North Devon.

Since relocating to North Devon in 2005 we have become ever more closely involved with other community organisations, and we hope that this trend will continue. We have also worked closely with Police Forces across the UK as consultants for animal mutilation cases, and we intend to forge closer links with the coastguard and other community services. We want to work closely with those who regularly travel into the Bristol Channel, so that if the recent trend of exotic animal visitors to our coastal waters continues, we can be out there as soon as possible.

Apart from having been the only Fortean Zoological organisation in the world to have consistently published material on all aspects of the subject for over a decade, we have achieved the following concrete results:

- Disproved the myth relating to the headless so-called sea-serpent carcass of Durgan beach in Cornwall 1975
- Disproved the story

of the 1988 puma skull of Lustleigh Cleave

- Carried out the only in-depth research ever into the mythos of the Cornish Owlman.
- Made the first records of a tropical species of lamprey
- Made the first records of a luminous cave gnat larva in Thailand
- Discovered a possible new species of British mammal - the beech marten
- In 1994-6 carried out the first archival fortean zoological survey of Hong Kong
- In the year 2000, CFZ theories were confirmed when a new species of lizard was added to the British List
- Identified the monster of Martin Mere in Lancashire as a giant wels catfish
- Expanded the known range of Armitage's skink in the Gambia by 80%
- Obtained photographic evidence of the remains of Europe's largest known pike
- Carried out the first ever in-depth study of the ninki-nanka
- Carried out the first attempt to breed Puerto Rican cave snails in captivity
- Were the first European explorers to visit the `lost valley` in Sumatra
- Published the first ever evidence for a new tribe of pygmies in Guyana
- Published the first evidence for a new species of caiman in Guyana

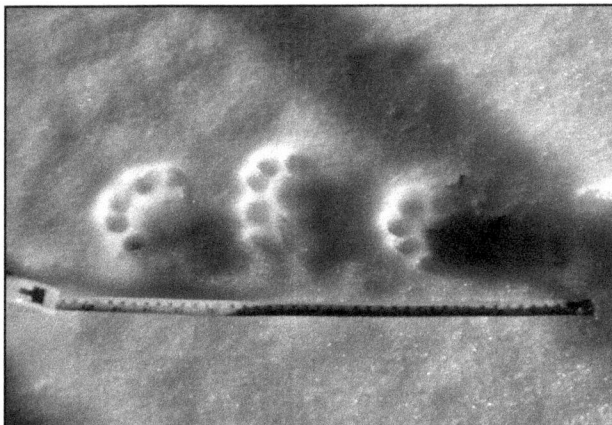

on a monster-haunted lake in Ireland for the first time
• Had a sighting of orang pendek in Sumatra in 2009
• Found leopard hair, subsequently identified by DNA analysis, from rural North Devon in 2010
• Brought back hairs which appear to be from an unknown primate in Sumatra
• Published some of the best evidence ever for the almasty in southern Russia

CFZ Expeditions and Investigations include:

• 1998 Puerto Rico, Florida, Mexico (Chupacabras)
• 1999 Nevada (Bigfoot)
• 2000 Thailand (Naga)
• 2002 Martin Mere (Giant catfish)
• 2002 Cleveland (Wallaby mutilation)
• 2003 Bolam Lake (BHM Reports)

- 2003 Sumatra (Orang Pendek)
- 2003 Texas (Bigfoot; giant snapping turtles)
- 2004 Sumatra (Orang Pendek; cigau, a sabre-toothed cat)
- 2004 Illinois (Black panthers; cicada swarm)
- 2004 Texas (Mystery blue dog)
- Loch Morar (Monster)
- 2004 Puerto Rico (Chupacabras; carnivorous cave snails)
- 2005 Belize (Affiliate expedition for hairy dwarfs)
- 2005 Loch Ness (Monster)
- 2005 Mongolia (Allghoi Khorkhoi aka Mongolian death worm)

- 2006 Gambia (Gambo - Gambian sea monster , Ninki Nanka and Armitage's skink
- 2006 Llangorse Lake (Giant pike, giant eels)
- 2006 Windermere (Giant eels)
- 2007 Coniston Water (Giant eels)
- 2007 Guyana (Giant anaconda, didi, water tiger)
- 2008 Russia (Almasty)
- 2009 Sumatra (Orang pendek)
- 2009 Republic of Ireland (Lake Monster)
- 2010 Texas (Blue Dogs)
- 2010 India (Mande Burung)
- 2011 Sumatra (Orang-pendek)

For details of current membership fees, current expeditions and investigations, and voluntary posts within the CFZ that need your help, please do not hesitate to contact us.

The Centre for Fortean Zoology,
Myrtle Cottage,
Woolfardisworthy,
Bideford, North Devon
EX39 5QR

Telephone 01237 431413
Fax+44 (0)7006-074-925
eMail info@cfz.org.uk

Websites:

www.cfz.org.uk
www.weirdweekend.org

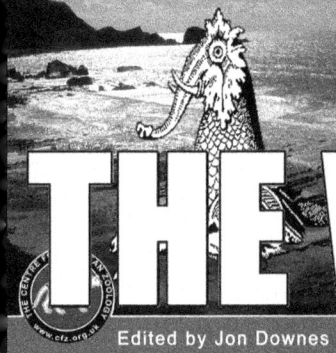

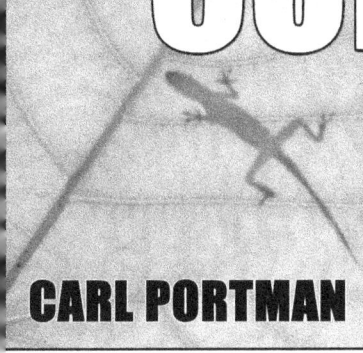

THE WORLD'S WEIRDEST PUBLISHING COMPANY

ANIMALS & MEN ISSUES 16-20
THE JOURNAL OF THE CENTRE FOR FORTEAN ZOOLOGY
NEW HORIZONS
Edited by Jon Downes
www.cfz.org.uk

BIG CATS
LOOSE IN BRITAIN

PREDATOR DEATHMATCH
NICK MOLLOY
WITH ILLUSTRATIONS BY ANTHONY WALLIS

Edited by
Jonathan Downes and Richard Freeman

FOREWORD BY Dr. KARL SHUKER

A DAINTREE DIARY
Tales from Travels
tropical North Daintree
CARL PORTMAN

THE COLLECTED POEMS
Dr Karl P.N. Shuker

STRANGELYSTRANGE
ly normal
an anthology of writings by
ANDY ROBERTS

HOW TO START A PUBLISHING EMPIRE

Unlike most mainstream publishers, we have a non-commercial remit, and our mission statement claims that "we publish books because they deserve to be published, not because we think that we can make money out of them". Our motto is the Latin Tag *Pro bona causa facimus* (we do it for good reason), a slogan taken from a children's book *The Case of the Silver Egg* by the late Desmond Skirrow.

WIKIPEDIA: "The first book published was in 1988. *Take this Brother may it Serve you Well* was a guide to Beatles bootlegs by Jonathan Downes. It sold quite well, but was hampered by very poor production values, being photocopied, and held together by a plastic clip binder. In 1988 A5 clip binders were hard to get hold of, so the publishers took A4 binders and cut them in half with a hacksaw. It now reaches surprisingly high prices second hand.

The production quality improved slightly over the years, and after 1999 all the books produced were ringbound with laminated colour covers. In 2004, however, they signed an agreement with Lightning Source, and all books are now produced perfect bound, with full colour covers."

Until 2010 all our books, the majority of which are/were on the subject of mystery animals and allied disciplines, were published by `CFZ Press`, the publishing arm of the Centre for Fortean Zoology (CFZ), and we urged our readers and followers to draw a discreet veil over the books that we published that were completely off topic to the CFZ.

However, in 2010 we decided that enough was enough and launched a second imprint, `Fortean Words` which aims to cover a wide range of non animal-related esoteric subjects. Other imprints will be launched as and when we feel like it, however the basic ethos of the company remains the same: Our job is to publish books and magazines that we feel are worth publishing, whether or not they are going to sell. Money is, after all - as my dear old Mama once told me - a rather vulgar subject, and she would be rolling in her grave if she thought that her eldest son was somehow in `trade`.

Luckily, so far our tastes have turned out not to be that rarified after all, and we have sold far more books than anyone ever thought that we would, so there is a moral in there somewhere…

Jon Downes,
Woolsery, North Devon
July 2010

CFZ PRESS

Other Books in Print

Wildman! by Redfern, Nick
Globsters by Newton, Michael
Cats of Magic, Mythology and Mystery Shuker, by Karl P. N
Those Amazing Newfoundland Dogs by Bondeson, Jan
The Mystery Animals of Pennsylvania by Gable, Andrew
Sea Serpent Carcasses - Scotland from the Stronsa Monster to Loch Ness by Glen Vaudrey
The CFZ Yearbook 2012 edited by Jonathan and Corinna Downes
ORANG PENDEK: Sumatra's Forgotten Ape by Richard Freeman
THE MYSTERY ANIMALS OF THE BRITISH ISLES: London by Neil Arnold
CFZ EXPEDITION REPORT: India 2010 by Richard Freeman *et al*
The Cryptid Creatures of Florida by Scott Marlow
Dead of Night by Lee Walker
The Mystery Animals of the British Isles: The Northern Isles by Glen Vaudrey
THE MYSTERY ANIMALS OF THE BRTISH ISLES: Gloucestershire and Worcestershire by Paul Williams
When Bigfoot Attacks by Michael Newton
Weird Waters – The Mystery Animals of Scandinavia: Lake and Sea Monsters by Lars Thomas
The Inhumanoids by Barton Nunnelly
Monstrum! A Wizard's Tale by Tony "Doc" Shiels
CFZ Yearbook 2011 edited by Jonathan Downes
Karl Shuker's Alien Zoo by Shuker, Dr Karl P.N
Tetrapod Zoology Book One by Naish, Dr Darren
The Mystery Animals of Ireland by Gary Cunningham and Ronan Coghlan
Monsters of Texas by Gerhard, Ken
The Great Yokai Encyclopaedia by Freeman, Richard
NEW HORIZONS: Animals & Men issues 16-20 Collected Editions Vol. 4 by Downes, Jonathan
A Daintree Diary -
Tales from Travels to the Daintree Rainforest in tropical north Queensland, Australia by Portman, Carl
Strangely Strange but Oddly Normal by Roberts, Andy

by Downes, Jonathan
The Smaller Mystery Carnivores of the Westcountry by Downes, Jonathan
CFZ EXPEDITION REPORT: Gambia 2006 by Richard Freeman *et al*, Shuker, Karl (fwd)
The Owlman and Others by Jonathan Downes
The Blackdown Mystery by Downes, Jonathan
Big Cats in Britain Yearbook 2006 by Fraser, Mark (Ed)
Fragrant Harbours - Distant Rivers by Downes, John T
Only Fools and Goatsuckers by Downes, Jonathan
Monster of the Mere by Jonathan Downes
Dragons:More than a Myth by Freeman, Richard Alan
Granfer's Bible Stories by Downes, John Tweddell
Monster Hunter by Downes, Jonathan

CFZ Classics is a new venture for us. There are many seminal works that are either unavailable today, or not available with the production values which we would like to see. So, following the old adage that if you want to get something done do it yourself, this is exactly what we have done.

Desiderius Erasmus Roterodamus (b. October 18th 1466, d. July 2nd 1536) said: "When I have a little money, I buy books; and if I have any left, I buy food and clothes," and we are much the same. Only, we are in the lucky position of being able to share our books with the wider world. CFZ Classics is a conduit through which we cannot just re-issue titles which we feel still have much to offer the cryptozoological and Fortean research communities of the 21st Century, but we are adding footnotes, supplementary essays, and other material where we deem it appropriate.

Headhunters of The Amazon by Fritz W Up de Graff (1902)

Fortean Words

The Centre for Fortean Zoology has for several years led the field in Fortean publishing. CFZ Press is the only publishing company specialising in books on monsters and mystery animals. CFZ Press has published more books on this subject than any other company in history and has attracted such well known authors as Andy Roberts, Nick Redfern, Michael Newton, Dr Karl Shuker, Neil Arnold, Dr Darren Naish, Jon Downes, Ken Gerhard and Richard Freeman.

Now CFZ Press are launching a new imprint. Fortean Words is a new line of books dealing with Fortean subjects other than cryptozoology, which is - after all - the subject the CFZ are best known for. Fortean Words is being launched with a spectacular multi-volume series called *Haunted Skies* which covers British UFO sightings between 1940 and 2010. Former policeman John Hanson and his long-suffering partner Dawn Holloway have compiled a peerless library of sighting reports, many that have not been made public before.

Other books include a look at the Berwyn Mountains UFO case by renowned Fortean Andy Roberts and a series of forthcoming books by transatlantic researcher Nick Redfern. CFZ Press are dedicated to maintaining the fine quality of their works with Fortean Words. New authors tackling new subjects will always be encouraged, and we hope that our books will continue to be as ground-breaking and popular as ever.

Haunted Skies Volume One 1940-1959 by John Hanson and Dawn Holloway
Haunted Skies Volume Two 1960-1965 by John Hanson and Dawn Holloway
Haunted Skies Volume Three 1965-1967 by John Hanson and Dawn Holloway
Haunted Skies Volume Four 1968-1971 by John Hanson and Dawn Holloway
Haunted Skies Volume Five 1972-1974 by John Hanson and Dawn Holloway
Haunted Skies Volume Six 1975-1977 by John Hanson and Dawn Holloway
Grave Concerns by Kai Roberts

Police and the Paranormal by Andy Owens
Dead of Night by Lee Walker
Space Girl Dead on Spaghetti Junction - an anthology by Nick Redfern
I Fort the Lore - an anthology by Paul Screeton
UFO Down - the Berwyn Mountains UFO Crash by Andy Roberts
The Grail by Ronan Coghlan
UFO Warminster - Cradle of Contract by Kevin Goodman
Quest for the Hexham Heads by Paul Screeton

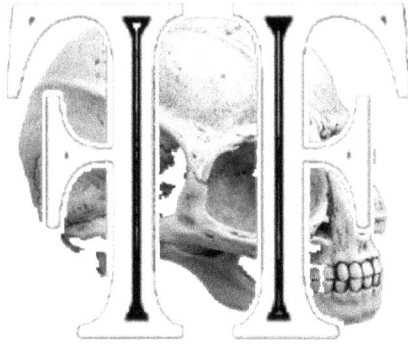

Fortean Fiction

J ust before Christmas 2011, we launched our third imprint, this time dedicated to - let's see if you guessed it from the title - fictional books with a Fortean or cryptozoological theme. We have published a few fictional books in the past, but now think that because of our rising reputation as publishers of quality Forteana, that a dedicated fiction imprint was the order of the day.

We launched with four titles:

Green Unpleasant Land by Richard Freeman
Left Behind by Harriet Wadham
Dark Ness by Tabitca Cope
Snap! By Steven Bredice
Dark Wear by Tabitca Cope
Hyakymonogatari Book 1 by Richard Freeman

www.ingramcontent.com/pod-product-compliance
Lightning Source LLC
Chambersburg PA
CBHW062208270326
41930CB00009B/1686